A Dictionary of Environmental Economics, Science, and Policy

R. Quentin Grafton
Institute of the Environment and Department of Economics, University of Ottawa, Canada

Linwood H. Pendleton
School of International Relations and Wrigley Institute for Environmental Studies, University of Southern California, USA

Harry W. Nelson
Forest Economics and Policy Analysis Research Unit, University of British Columbia, Canada

Edward Elgar
Cheltenham, UK • Northampton, MA, USA

Published by
Edward Elgar Publishing Limited
Glensanda House
Montpellier Parade
Cheltenham
Glos GL50 1UA
UK

Edward Elgar Publishing, Inc.
136 West Street
Suite 202
Northampton
Massachusetts 01060
USA

A catalogue record for this book
is available from the British Library

Library of Congress Cataloguing in Publication Data
Grafton, R. Quentin, 1962–
    A dictionary of environmental economics, science and policy / R. Quentin
    Grafton, Linwood H. Pendleton, Harry W. Nelson.
        p. cm.
        1. Environmental economics—Dictionaries. 2. Environmental
        science—Dictionaries. I. Pendleton, Linwood H., 1964– II. Nelson,
        Harry W., 1961– III. Title.
    HC79.E5 G685 2001
    333.7'03—dc21

ISBN 1 84064 126 6

Printed and bound in Great Britain by MPG Books Ltd, Bodmin, Cornwall

A DICTIONARY OF ENVIRONMENTAL ECONOMICS, SCIENCE, AND POLICY

For Ariana and Brecon, Emma, and Alexa

Table of Contents

Acknowledgements

We are very grateful for the many authors whose papers and books helped us to understand environmental economics, science, and policy, so that we could help others.

We very much appreciate the research assistance of Josée Benoît, D.G. Webster, and Cheryl Cuesta. We also thank David J. Leech for helping us to prepare the text in the format required by the publishers. Merci beaucoup!

Our thanks also to Dymphna Evans at Edward Elgar for her encouragement, and to Jeroen van den Bergh and Pam Mason for their most helpful reviews.

Preface

With the increasing scarcity of environmental services and natural resources, it has become ever more critical to manage the environment more efficiently. Resource managers and environmental regulators must combine natural science and economics into more effective policies. The natural sciences are critical for understanding the link between actions and consequences in nature. Economics is critical for understanding what markets will do and how to design public programs to maximize net social benefits.

As most observers instinctively know, the economy is the greatest threat to nature. It is economic activities that are at the heart of most pollution, most solid and municipal waste, and most of the destruction of natural systems. In society's active pursuit of material wealth, the environment can sometimes suffer. One reason to study economics is to better understand the economy and understand why the economy uses natural resources, pollutes the air and water, and compromises pristine ecosystems.

There are many reasons why markets may not make efficient environmental and natural resource decisions. Sometimes market decision makers simply overlook their impact on the environment. Sometimes there are multiple owners of resources and responsibility is diluted. Sometimes there are no owners of a resource and people rush to use it before anyone else can. All of the above market failures suggest that the economy sometimes fails to manage environmental and natural resources carefully.

When market failures occur, the government must play an active role, regulating the market and encouraging investments into natural resources that would improve social outcomes. Economics plays a critical role here identifying where governments need to get involved and where markets serve effectively by themselves. Economics also helps identify how governments can best encourage markets to become efficient, how taxes can be more effectively used to control pollution, when private property rights are helpful, how trade should be managed to protect the environment, and when the government should directly manage resources.

An important second role for economics is in providing a framework to help analyze and design more effective social programs and regulations. This framework, cost–benefit analysis, encourages governments to design programs efficiently. Environmental and natural resource programs should maximize net social benefits, the amount that benefit exceed costs. This requires governments to carefully examine and measure all the benefits of taking different actions and to make sure that these benefits are large. The government must also consider all costs and try to keep these costs well below benefits. Helping design better environmental and natural resource policies is an important role for economics.

Introducing economics into natural science and management programs and introducing natural science to economics programs has been a challenge for many traditional educational programs. Each discipline has its own language and terms. Although these terms are helpful in developing a careful discourse in each field, they also serve as barriers for interdisciplinary efforts. This technical dictionary serves an important need in many interdisciplinary programs. By providing definitions of many key economic terms, the dictionary can help students and practising professionals with diverse backgrounds master the language of economics and environmental sciences more easily. The dictionary will serve a very useful role, helping people interested in environmental and natural resource topics understand each other.

The book serves another important role as well. By providing a basic primer on economics, international environmental problems and ecological modeling, it provides a wonderful introduction to environmental–economic models and thinking. The first primer can give the reader a quick grasp of the most critical insights that economics has to offer environmental management without being waylaid by an entire text. The second primer reviews the major conflicts arising in the international environmental arena. The third primer covers a set of ecosystem models from basic nutrient flow diagrams to fish populations. These primers serve as excellent introductions to each field.

The final contribution of the book is an excellent set of appendices that are basic reference materials in this field. The appendices cover the Greek alphabet (used often in modeling), Roman numerals, international scientific units, international prefixes, abbreviations, geological time, and international environmental treaties. This is a highly useful reference section for economics, international law and environmental science.

This book is a valuable reference source for professionals in natural resource and environmental management and for non-specialists. I recommend that practitioners get a copy for their own reference and that courses in this area adopt the book as a supplemental text.

Robert Mendelsohn
Edwin Weyerhaeuser Davis Professor
Yale University

Introduction

Without a holistic view of the environment that transcends disciplines, we can never hope to have a comprehensive understanding of our world. Unfortunately, the language of each discipline includes its own set of jargon and words that are commonly used by people within a discipline but, frequently, cannot be understood by people from the outside. Too often, these barriers prevent us from understanding and developing a systematic view of our environment.

A Dictionary of Environmental Economics, Science, and Policy offers a reference that bridges the gap between the disciplines of environmental economics, environmental sciences and environmental studies. It provides a comprehensive set of words used in environmental, ecological and resource economics. In addition, the dictionary includes a selection of some of the most important and frequently used words in the environmental sciences and environmental studies. The book includes over 3,300 words from *abatement* to *zooplankton* and, where further explanations are required, further reading and references are provided. The dictionary is not only a ready-reference for students, but should prove useful to policy makers and professionals who need to understand the many different terms and concepts about the environment.

The book has a number of unique features. Included are three primers that provide introductions to the topics of Environmental Economics, International Environmental Problems, and Environmental Systems, Dynamics and Modeling. These primers can be used as reviews or introductions in many different courses and programs on the environment. In addition, the dictionary has a list of annotated references that provides a useful introduction to the topics covered in the dictionary, a comprehensive list of references referred to in the definitions and seven appendices that include: the Greek alphabet, Roman numerals, the système internationale (SI) units, prefixes for the SI, common abbreviations for measures and units, the geological time-scale, and a listing and description of selected environmental treaties. The appendix on environmental treaties should be helpful to readers who are interested in knowing what we are doing to sustain the global environment.

We trust that our book, with its primers and appendices, will prove a ready-reference that will bridge environmental disciplines and will help you to better understand the world we share.

R.Q.G., Ottawa, Canada
L.H.P., Los Angeles, USA
H.W.N., Vancouver, Canada

Economics for the Environment: A Primer

1. WELFARE ECONOMICS AND MARGINAL ANALYSIS

Much of what we know as environmental economics can best be thought of as applied economics. The positive side of environmental economics uses micro and macroeconomic theory to learn more about the ways in which economic factors influence the consumption of environmental goods and services. Positive environmental economics is largely descriptive and predictive. The normative aspect of environmental economics is largely prescriptive and applies principles of welfare economics to determine the allocation of environmental goods and services that generates the greatest economic good for society.

Normative welfare economics provides a framework to determine how society "should" allocate resources, assuming that economic welfare is a good measure of society's well-being. **Pareto** pointed out that an optimal allocation of goods exists when any reallocation would only make one person better off at the expense of another; we call this allocation a **Pareto efficient** or **economically efficient** allocation. At sub-optimal allocations, we could potentially reallocate resources to make everyone better off as whole. Reallocations away from the sub-optimal towards the optimal are known as **Pareto improvements**. Using the ideas of Pareto, normative environmental economics provides a framework for finding the economically optimal allocation of environmental resources.

Talking about the greatest good to society and measuring this good are quite different things. To make discussions of well-being more tangible, economists need to find concrete ways of capturing the values people give to goods, resources, and even services. Generally, we say that someone values a good if they are willing to give up something they already have or are entitled to have to obtain the good in question. In subsistence economies, people may trade their time (say time spent hunting) in order to get things they value (such as game). In barter systems, people trade among each other by offering one good for another. In market systems, we trade currency (whether be it Euros or Treasury Notes) for goods. The reasons people are willing to make sacrifices to get goods or services may be complex. Nevertheless, the fact that people are willing to make tangible trades gives us clues about people's valuations and choices.

a. Measuring the Value of Goods

The market price of a good is an important indication of the value a consumer places on that good. Since people presumably do not value currency in and of itself, economists accept that the amount paid in currency is a simple reflection of the **current value** a consumer places on a good. When consumers hand over 50 cents to buy an apple, they reveal that they place a value on that apple of at least 50 cents. Even when goods are sold individually it is common for consumers to buy more than one unit of a good, either at once or over the course of some period of time. The total number of units bought depends on the price; when prices are high consumers buy less of a good and when prices are low consumers buy more. As before, the consumer only buys apples as long as the value he or she places on each apple exceeds the price per apple in the market. Of course, the consumer never buys an infinite number of apples, so we know immediately that the value a consumer places on any individual apple must fall as that consumer acquires more and more apples. We call the value that consumers place on each newly purchased apple their **marginal willingness to pay** for that last apple. The consumer will keep buying a good as long as his/her **marginal willingness to pay** for the good is at least equal to the price of the good. Therefore, if a consumer buys 10 apples at 5 cents/apple we know that the value of the last apple is 5 cents/apple. Does that mean the value of all 10 apples is 5 cents times 10 or 50 cents? Not necessarily. To understand why, we need to think about why a consumer buys more when the price is low and less when the price is high. We call the study of this relationship of prices and quantities "demand analysis" and the relationship a **demand function**.

b. Demand

In the market, we observe regular patterns in which the quantity of a good purchased by a consumer varies with the price of the good. We call the quantity of a good that a consumer would like to buy for a given price, the demand for that good. When prices are high, demand is small, and vice-versa. While prices usually determine demand, we have already shown that when a consumer makes a purchase, the purchase price reveals the value that the consumer places on a good. A consumer places a higher value on a single apple than on one of 10 apples, especially if the 10th apple is consumed after the first 9 have been eaten.

We call the schedule of quantities demanded at different prices the **Marshallian demand** function or simply the demand curve for a good.

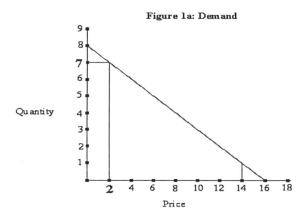

Figure 1a: Demand

The **demand function** reveals how much of a good a consumer would buy at a given price (e.g., when the price is 2, the demand is 7). Figure 1a shows that at a price (drawn on the horizontal axis) of 0, the consumer would demand (drawn on the vertical axis) 8 units and at a price of 16 the demand falls to 0 (we call this the **choke price**). The actual schedule of prices and quantities demanded depends on the consumer's personal characteristics, including income. For instance, rich people may buy more of a good at a given price than poor people. The slope and the shape of the demand curve (a demand curve can bend in towards the origin) depends on such things as how essential the good is or how many substitutes for the good are available to the consumer.

If we switch the axes of our graph, we now have a schedule that reveals the value a consumer places on each unit of a good, depending on how many units already have been consumed. As before, we call the value that is placed on one more unit of a good, the **marginal**

willingness to pay for that unit, and the new schedule of prices as a function of quantity the **inverse demand function or curve**.

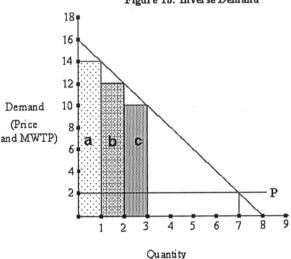

Figure 1b: Inverse Demand

Once again, when 7 units are purchased, the 7th unit is valued at 2. Note that the 8th unit has zero value, that is, the consumer would consume up to 8 units if they were free, but any more than that would not make the consumer any better off.

The inverse demand curve also reveals both the value of the last marginal unit (**the marginal willingness to pay**) and also the cumulative value that the consumer would place on the consumption of all the units consumed (**the total willingness to pay**). The marginal willingness to pay is simply read off the inverse demand curve. The total willingness to pay, however, must be found by summing together the **marginal willingness to pay** for every individual unit. For example, the total willingness to pay for 3 units would be equal to the marginal willingness to pay for each unit separately (Figure 1b, areas a+b+c). A continuous inverse demand function, however, shows that the **marginal willingness to pay** varies for even small fractions of a unit. The stair steps of areas a, b and c in Figure 1b only approximate the **total willingness to pay** for 3 units of the good. The marginal willingness to pay is an exact measure only of the infinitely small fraction of the last unit consumed. If we make the units smaller and smaller, the exact **total willingness to pay** is found by summing up increasingly tiny slivers under the demand curve. The summation of these slivers equals the area under the inverse demand curve up to the quantity consumed.[v] In Figure 1c, we draw the **total willingness to pay** for 3 units.

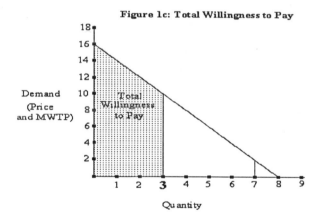

While total willingness to pay represents the gross value a consumer places on the consumption of a good, economists generally are interested

---

[v] Readers familiar with calculus may recognize that the total willingness to pay is simply the integral of the marginal willingness to pay evaluated from 0 to the quantity consumed. Conversely, marginal willingness to pay is the derivative of total willingness to pay.

in the benefit of a good or resource net of the cost of obtaining that good. From the consumers' perspective, the cost of a good is the total price paid for a good (price × quantity). If only a single, uniform price is charged for each unit of a good, then the total willingness to pay for a good will exceed the cost to the consumer (price paid). The difference between total willingness to pay and the actual amount paid is called **consumer surplus** (Figure 1d).

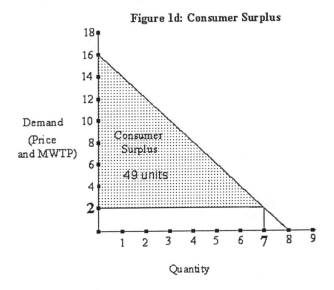

Figure 1d: Consumer Surplus

This is the most basic measure of the contribution of a good to a consumer's well-being. Other more precise measures of welfare include the compensated welfare measures (e.g., **compensating variation** and **equivalent variation**).

c. Aggregate Demand and Society's Welfare

The leap from measuring the welfare of a consumer to measuring the welfare of society requires some sort of aggregation of values across people. If we examine the problem strictly from the perspective of market demand, we can simply add together the demand that each consumer expresses for a good at the given price. For instance, if at a price of 2 cents, José demands 5 apples and Sue wants 3 apples, then their combined demand is simply 8 apples when the price is 2 cents. By adding up across individual consumers, we can derive a market demand function that has all of the properties we described above (e.g., inverse demand still reflects marginal willingness to pay, but now for all consumers). In a similar manner, we calculate **consumer surplus** for all consumers that participate in the market.

d. Supply

An economy consists of more than consumers; we also need to look at the supply of goods and the costs of providing goods. Supply analysis can be graphically demonstrated using the same marginal framework we developed for demand. The cost of providing one more good is the marginal cost of that good. As before, we draw the quantity of a good provided on the horizontal axis. The cost of providing one more unit of a good, its marginal cost is drawn on the vertical axis. The relationship between quantity and marginal cost is called the supply function or the marginal cost function. Marginal cost may increase, remain constant, decrease, or even develop an irregular shape known as a **backward bending supply curve**. The total cost of supplying a good is found by determining the area under the supply curve between zero and the quantity supplied.

Producers will supply goods as long as the price on the market is greater than the marginal cost of production. If the supply curve is upward sloping, then the price will exceed the marginal cost for every unit but the last and the producer will earn a benefit known as **rent** or **producer surplus**. Figure 2 demonstrates an upward sloping supply function, price, and the associated producer surplus that comes from supplying goods up to the point where marginal cost equals price. In the aggregate, there exists a market supply function that reflects the marginal costs of supplying goods when all producers are considered. Some producers may be able to produce goods at low marginal costs and others may produce at higher marginal costs. Those that can produce goods

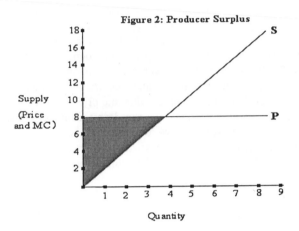

Figure 2: Producer Surplus

more cheaply will earn more benefits and thus more **rent** or **producer surplus**.

2. MARKET EQUILIBRIUM AND MARKET FAILURE

In market economies, prices are not set arbitrarily, but are the outcome of the forces of supply and demand. Producers will supply goods as long as consumers are willing to pay at least the marginal costs of supply. Consumers will buy goods as long as their marginal willingness to pay (marginal benefit) exceeds the costs charged by producers. Through market processes of bids and offers, consumers and producers make trades. If consumers demand more goods than a given set of producers can provide, new producers will enter the market as long as they can supply goods at a cost less than the consumers' willingness to pay. As more and more goods are produced and consumed, two things may happen: (1) because of the property of declining marginal benefits, the consumers' willingness to pay will fall and (2) the marginal costs of production may remain constant or increase, perhaps due to technological reasons. Eventually market trades reach a point where the marginal cost of production equals the marginal benefit to consumers. We call this an equilibrium point (demonstrated in Figure 3a) where q* is the equilibrium allocation and p* is the equilibrium price.

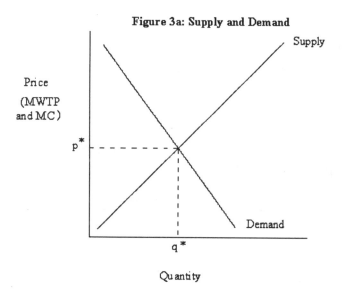

Figure 3a: Supply and Demand

If the amount supplied is less than q*, a producer can earn rent by increasing the supply. If more than the equilibrium quantity is supplied, consumers will not pay a price sufficient to cover marginal costs and some producers will lose money and reduce their supply.

As long as market costs reflect the actual cost of goods, consumers and producers will trade at an allocation that maximizes welfare to society, where welfare is **consumer surplus** plus **producer surplus** or **net social surplus** (Figure 3b).

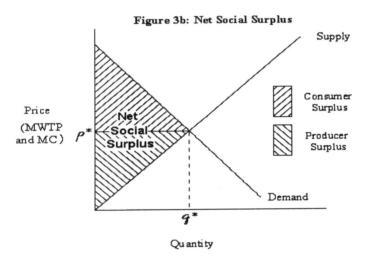

Figure 3b: Net Social Surplus

(When net social surplus is maximized, we say we have an **economically efficient** allocation.) This welfare maximizing property of free markets has led many to support market-based mechanisms for the allocation of resources (**market environmentalists**), and has given rise to a movement known as Free Market Environmentalism. The conditions under which a market must operate to achieve the maximum welfare for society are, however, difficult to find. At the minimum, a well functioning market needs to be one in which **property rights** for resources are well defined, there are many buyers and sellers, and prices most accurately represent the costs of providing goods. Frequently these conditions are not met and we say the economy experiences **market failure**.

The efficient allocation of natural resources and environmental goods is routinely impeded by **market failures**. Property rights do not exist for many resources (e.g., deep sea minerals or open ocean fisheries). If resources are open to all (**open-access** resources), the market will not allocate these resources in a way that maximizes the benefits to society. Sometimes the lack of property rights stems from a difficulty in laying claim to **fugitive,** mobile and remote resources. When this is the case, steps can be taken to assign property rights, thus improving the degree to which the market functions properly. In other cases, resources lack "owners" because their very nature defies ownership. When consumers cannot be excluded from enjoying a resource and when the enjoyment of that resource by one consumer does not diminish the enjoyment of that resource by another, we call the good in question a **public good**. Examples of **public goods** include national defense, air quality, and whale conservation (public bads include things like global warming and coastal pollution). Because property rights cannot be assigned to public goods, the market is unlikely to allocate public good resources in a way that is **economically efficient** (i.e., the welfare maximizing allocation is not chosen by the market).

Often environmental impacts are the byproduct of other activities. Recycling trucks create noise when they drive through your neighborhood to pick up bottles; power plants emit smoke and steam that can obstruct views; surfers can create entertaining viewing for beachgoers. In all three cases, we find that the people who bear the costs or enjoy the benefits of the "byproducts" of these activities may be different than the people whose activities create these "byproducts". When costs and benefits of an activity are not borne by the immediate consumer or producer, we call these costs "**negative externalities**" and the benefits "**positive externalities**". As in the examples of open access and public goods, the problem is that property rights to the externalities have not been properly defined. When property rights for an externality are well defined (e.g., people have property rights to be free from noise or recycling trucks have the right to make noise), we expect a well functioning market to allocate resources efficiently (see **Coase Theorem**). Property rights can sometimes be assigned directly to externalities. In other cases, government can internalize externalities by determining the economic value of an impact and directly charging those responsible for negative externalities (see, for example, **Pigouvian taxes**) or by subsidizing positive externalities.

3. NON-MARKET VALUES AND TOTAL ECONOMIC VALUE

Many environmental and natural resources are not traded in markets, either because they are public goods or because property rights to these goods have not been properly assigned. Unlike most standard market goods (say a candy bar), certain environmental goods may possess values that can be considered as **public goods** and other values that are purely **private** in nature. It would be hard to imagine how a candy bar could be anything other than a **private good**. On the other hand, consider a tree. When a tree is cut for timber, it is consumed as a **private good**. That same tree, however, could also have been viewed by hundreds of hikers had it remained alive in the forest. Viewing that tree would represent a **public good** value associated with the tree. The fact that natural and environmental resources tend to have properties that are both public and private, utilitarian and spiritual, practical and intangible means these goods are far more complex than standard market goods. To understand when and why markets allocate and misallocate environmental and natural resources, we need to start by understanding the types of values these environmental goods can possess.

To start, we break down the **total economic value** of an environmental good or service into those values that people derive from direct use (called **use values**) and those values that do not require direct use by the consumer (**non-use** or **passive use values**). **Use values** include all of those transitive, active uses of a good or resource. Some use values have properties of private goods. These values typically involve some kind of extraction or consumption that allows the consumer to exclude others from using the good or resource. Timbering a tree is one example, but many less dramatic examples also exist including picking up seashells or diverting water from a lake to irrigate your garden. When a good is removed from its place in nature and consumed in a way that makes exclusion possible, we call the value associated with that good an **extractive use** value. If environmental goods had only extractive use values, we would expect that the market might be able to allocate these goods in an **economically efficient** way. However, the very same goods may also have **non-extractive use values**. **Non-extractive use values** are those values that emanate from the use of a good that do not require removal from nature and subsequently are difficult to exclude from use by others. Viewing a tree is a **non-extractive use value**, wading in the ocean, and sailing on a lake are both activities that generate non-extractive use values.

Environmental goods differ from other common market goods in that many environmental goods possess **non-use** or **passive use values**. Non-use values are values that a consumer derives from a good that they never use or do not plan to use directly. For instance, some people may derive utility just from knowing that sea otters exist somewhere in California (we call this **existence value**). These people may even be willing to pay to protect these animals (direct evidence of the minimum value they place on this passive use). Other **passive use values** include **bequest value** and **option value**. Whether **non-use values** are true values, distinct from some sort of expected use value, is a matter of some debate. Furthermore, the universal acceptance of **non-use** values by many policy makers is hindered by the difficulty economists have had in finding reliable techniques to measure these values. Nevertheless, a "blue ribbon" panel convened by the National Oceanic and Atmospheric Agency of the United States (**NOAA panel**) found that **non-use values** were at least sufficiently legitimate to be used as a starting point for assessing economic losses in a judicial setting.

4. MEASURING NON-MARKET VALUES

Measuring the non-market values of environmental goods has been one of the greatest challenges for environmental economics during the last 30 years. To capture the very real value of these goods, environmental economists have undertaken two avenues of empirical elicitation that can loosely be divided into "revealed preference, or, non-hypothetical approaches" and "hypothetical, or, contingent market" approaches. (As the science advances, these two approaches continue to blend together, but we treat them separately here to get several key points across.)

By definition, prices for "non-market" goods simply do not exist. Nevertheless, environmental values do influence market activities by influencing prices for certain goods (as **complements** or **substitutes** for market goods) and by influencing spending on other market goods. For instance, all things being equal, houses where the air is clean have higher market prices than houses in smoggy neighborhoods. When the quantity or quality of environmental goods and services affects consumer behavior we often can tease out the economic "value" of the environmental good through **econometrics**. We say that people "reveal their preferences" for environmental goods by their consumption behavior. Statistical methods can then be applied to econometric models of consumption that account for the ways in which environmental goods influence consumers' preferences for other goods. The class of methods that employs empirical statistical analysis of consumption behavior with models of consumption is known as the class of **revealed preference** methods. In the field of environmental economics, standard **travel cost methods**, **random utility models** and **hedonic methods** are three important revealed preference methods used to elicit the value of non-market environmental goods.

In many cases it is impossible to observe the ways in which environmental goods influence market goods. Sometimes, the environmental good of interest is independent of other market goods. Other times, the provision or loss of the good is rare or even entirely novel to the consumer – consider the loss of wildlife in Prince William Sound following the grounding of the Exxon Valdez or the hypothetical extinction of the Blue Whale. When the provision or loss of an environmental good is beyond our experience or falls outside of what we would normally consider complements or substitutes to market goods, we must turn to more hypothetical, or contingent, approaches to valuing environmental goods. The **contingent methods** use a variety of techniques to determine how consumers would value hypothetical changes in the quantity or quality of goods. The most basic of the contingent models use survey techniques to estimate the respondents' **willingness to pay** for the provision of an environmental good or the removal of an environmental bad, or their **willingness to accept** compensation for the opposite. Other contingent methods elicit values for environmental goods and services by having respondents rank environmental non-market goods with other market goods (**contingent ranking**). Increasingly, the contingent methods are designed to look more and more like market transactions. In the **stated preference method**, respondents are given a choice between hypothetical "bundles" of goods with each bundle containing at least one environmental good, and with hypothetical prices for every bundle. The consumer then chooses the bundle he or she prefers. In the stated preference approach, consumers' hypothetical choices can be treated much like the actual choices analyzed in the revealed preference methods. Finally, environmental economists can combine both revealed preference and stated preference approaches to fully utilize the full suite of available empirical and hypothetical data to provide clues about the values people give to non-market environmental goods.

5. SUSTAINABILITY

To address issues of sustainability, we must expand our discussion of resource allocation to include the idea that there may be an optimal way of distributing resources across time as well as among people. Under the umbrella of sustainability, we can consider two principal questions: (1) how long will our resources last, given current consumption patterns? and (2) how should we manage our resources so that future generations have access to the same quality of life as present generations?

The first question of resource longevity is really one of prediction and accounting; how do technology, taste, population, and natural regeneration influence the stock of environmental goods that will be available from one year to the next? Questions about how long resources will last have long concerned planners and economists who have worried that the world is running out of agricultural land, fish, and other important environmental goods. While **scarcity** may simply be defined as a quantity below which a good has a price in the market, the topic of **sustainability** attempts to frame the discussion as the degree to which a good is becoming more scarce. **Malthus** wrote that land for agriculture would become scarce, not because the quantity was disappearing but because the demand for land, driven by a growing population, was increasing dramatically. David Ricardo, on the other hand, pointed out that land may indeed become scarce, but not because it was in short supply. On the contrary, Ricardo believed land itself was abundant, but good, high quality land for agriculture was scarce and becoming scarcer for the same reasons proposed by Malthus. In either case, the concern was that the supply of a resource was not growing to keep pace with increasing demand.

A variety of indices have been developed to measure the degree to which resources are being consumed sustainably. In their most basic form, these indices simply divide stocks by consumption rates. More elaborate indices take into account the fact that changes in taste or technology may reduce future demand while natural regeneration, especially for biological resources, may increase stocks to keep pace with increasing demand. Generally, poor foresight has left predictions about resource sustainability wanting in accuracy. In a now famous bet, Professors Paul Ehrlich and Julian Simon wagered respectively that five important resource commodities were growing more scarce or less scarce. Ehrlich chose to bet that $1,000 invested equally in five metal commodities would increase in real value over a ten year period due to increasing scarcity. In fact, the real value of these commodities declined by more than fifty percent to $423.93. Clearly, a better understanding of trends in demand and technology are necessary before we can say with confidence how environmental quality and goods will fare in a consumption-based world.

The falling commodity prices shown in the wager between Ehrlich and Simon may have been misleading. Ten years is a short time frame.

What might have happened after 50 or 100 years? Also, market prices may signal market scarcity, but fail to indicate local scarcity. As markets grow and become global in scope, market signals for scarcity will also become global, revealing global conditions but masking local conditions of supply and demand.

The second question of sustainability "how should we manage our resources?" requires that we have some management target for the way in which resources are distributed across generations. The concept of **Pareto optimality**, useful in the analysis of welfare within a generation, is more difficult to apply when we do not know the tastes, preferences or technologies of future generations. Consumers tend to be impatient, preferring consumption this year over next (this is why we borrow). Are we to ignore this impatience when thinking of sustainability or should we include society's impatience by **discounting** the value of future consumption when we decide how to manage for sustainability? What about expected improvements in technology? Could saving too much of our natural and environmental resources make future generations better off at the expense of present generations?

The "we" in the second question of sustainability "how should *we* manage..." implies that there is some sort of shared vision about the goals of sustainability. While oft attempted, efforts to define a single target for sustainability have failed to achieve global consensus. Neoclassical economists often see the maximization of **net present value** as a metric against which to measure the sustainability of proposed resource use over time. Others have set out to define more broadly appealing, but less easily operable, definitions of sustainability.

The United Nations, under the mantle of the United Nations Commission on the Environment and Development (**UNCED**), turned to Gro Harlem Brundtland and a task force of experts (also known as the **Brundtland Commission**) to define sustainability. The **Brundtland Commission**, in turn, sought to frame the debate on sustainability in the context of economic development. While still vague in its prescriptions, the Brundtland Commission's positions on **sustainable development** made the trade-offs clear: economic growth should not come at the expense of future generations. Of course, determining what paths of resource use constitute a cost to the future still remains the topic of debate. However, whether sustainability and sustainable development ought to refer to the maintenance of the *status quo*, economic growth, redistribution of wealth, the protection of capital stocks or the preservation of **natural capital**, will vary by individual. Related to these concepts is the notion of **weak sustainability**. It supposes that we can substitute between human-made capital and natural capital in production and consumption, such that economic growth can be associated with improvements in environmental quality. By contrast, **strong sustainability** posits that natural and human-made capital are complements and cannot be substituted for each other in either production or consumption. Consequently, economic growth that uses natural resources and generates wastes must increase environmental degradation.

Regardless of the targeted goal of sustainability, the debate surrounding the topic has led to broad changes in the ways in which scholars think about resource use and the ways planners move forward with development projects. Today, a growing field of study known as **Ecological Economics** acts as a forum for exploring new ways to combine ecological knowledge with economic policy and analysis. In 1992, the United Nations hosted the Conference on Environment and Development in which many concepts of sustainability were formally institutionalized through an international declaration and a unified program of action.

The debate on **sustainability** and **sustainable development** is far from over. Few agree on the ability of technology to slow resource use or the ability of human-made capital to substitute for natural capital. For others, the debate centers on the fact that most programs designed to achieve a more sustainable world require substantial sacrifice that may not be shared equally among all. Nevertheless, the renewed focus on resource limits, first highlighted by Malthus and Ricardo, has led to the careful reconsideration of resource limits in models of development, growth and economic thought. In doing so, the scholarly and pragmatic pursuit of environmental understanding is increasingly realistic, more thorough, and more in tune with the ways in which economics, human society and the environment are linked.

International Environmental Problems: A Primer

One of the main virtues of the market system is that it allows people and firms to exchange goods at a relatively low cost by relying on a series of decentralized markets. The presence of these markets enables people to improve their welfare by exchanging the goods that they produce for goods that they can use and value more highly. The theory of international trade shows that the same principle is at work when countries engage in free trade: by specializing in the production of various goods in which they may enjoy either a **comparative advantage** or **absolute advantage** they are able to realize greater gains from trading goods with one another than simply consuming the goods themselves.

At the same time, many of the environmental problems we observe arise from some form of **market failure**. While markets may be efficient ways of distributing goods, they may not necessarily lead to optimal outcomes if the price of goods does not include all of the environmental costs associated with their production or consumption. One of the main ways in which all of the costs may not be included and markets might fail is when the rights to a resource are not well defined. Rights might not exist in the case of **open-access** or **common-pool resources**, where no one party owns the resource and the resource is free to be exploited by all, such as fisheries on the open sea. In this case, the price of the fish might not take into account the overexploitation of fish stocks and the risk of the stock falling below some critical level. Alternatively, the resource may be a **public good**, where the nature of the resource is such that it is impossible to exclude people from using or benefiting from it. Examples of such goods are the protective ozone layer in the earth's atmosphere or the genetic capital embodied in biodiversity. There may also be **public bads**, as in the case of greenhouse gases, when everybody potentially suffers the consequences of climate change. One consequence of public goods is that countries have incentives to let others bear the cost of any actions, such as maintaining biodiversity, reducing greenhouse gas emissions, or eliminating ozone-depleting substances, since they can enjoy the benefits from actions taken by other countries without incurring any cost themselves. This type of behavior is called **free-riding**, which can either lead to overexploitation of the resource (in the case of a public bad) or under-provision of the resource (in the case of a public good). Collectively, those common-pool resources and public goods that are truly international in nature are considered the **global**

commons: resources shared equally by all countries that can be potentially degraded or destroyed.

It may also be the case that the pollution or environmental degradation associated with the production or consumption of a good is not included in the price (**negative externalities**). Problems also arise when the scale of activity becomes great enough to have a measurable and negative impact on local ecosystems: for example, higher harvesting levels may reduce fish or wildlife populations below some critical level; or the cumulative effect from effluent flows discharged into a river from different countries may exceed the assimilative ability of a river to break down waste. Finally, problems can also occur from **policy failures**: government programs or regulations that have negative effects upon the environment. An example of such a failure might be subsidies to encourage fishers to switch from inshore fisheries to ocean fisheries, which increases pressure upon marine fish stocks.

Rapid population growth and subsequent development in countries around the world and the concomitant increase in trade have led to impacts outside the local environment and, increasingly, upon the global environment. At the same time, an increasing awareness of environmental issues and the interaction and interdependence of ecosystems, many of which cross national boundaries, has focused attention on the global or international nature of many of these problems. Solutions to environmental problems can be difficult in and of themselves: the nature of the environmental processes at work might not be well understood; it may be costly to identify the parties involved (either those who benefit or bear the costs); and it may be costly to create some kind of mechanism through which a solution can be implemented (whether it is through regulation, the allocation of rights, the establishment and collection of fees, or some other mechanism). These problems are only compounded when the problems cross political boundaries. The issues become particularly troublesome when the benefits and environmental costs are distributed such that some countries enjoy much of the benefit while others bear much of the environmental cost. Those bearing the damage may be countries downstream, or downwind, that receive the pollution from industrial activity in polluting countries that have little incentive to change their behavior if they do not suffer the consequences from their actions.

While there is no difference between the nature of the issues that arise in terms of global environmental problems and those that might arise in a community or country, there is one fundamental difference in terms of the environment in which they occur. International environmental problems, by definition, are those that involve more than one country;

consequently, there is no one overarching set of international laws or institutions that has the ability to address problems, enact solutions, and enforce them. To respond to those problems, countries have to pursue cooperative arrangements such as **Multilateral Environmental Agreements (MEAs)** which are agreements between different countries over how to handle various common environmental problems. As of today, there are hundreds of MEAs. A selection of these agreements is summarized in Appendix 7.

1. INTERNATIONAL AGREEMENTS

International agreements may take various forms. The most common are **treaties, conventions** and **protocols,** that are agreements between member countries. Treaties are legal agreements under international law in which the countries that sign the documents spell out their obligations or responsibilities. Conventions are agreements that establish a general framework or objectives, for which subsequent protocols will embody the details of how the objectives are to be achieved. Agreements do not become effective (that is, binding upon the countries that signed the agreement) until they come into force, and agreements do not come into force unless a sufficient number of countries have ratified the agreement. A treaty or convention only applies to those countries that are signatories, although provisions in the agreement may govern trade with non-parties, and countries may be able to make reservations at the time they sign an agreement (making a unilateral statement that modifies or excludes a certain portion of the agreement) as well as renounce their participation in an agreement at a later date.

The first set of international agreements between countries are primarily concerned with conflicts over shared resources, such as how to address the problems created by resources (e.g., fish stocks and migratory wildlife) that move between different countries or waterways that pass across national boundaries. In these cases, the **open-access** nature of the resource means that the first to capture the resource can lay claim to it, which can lead to overexploitation as countries "race for the resource". Examples of such agreements include the 1923 Convention for the Preservation of the Halibut Fishery of the Northern Pacific Ocean and the Bering Sea, and the treaties signed between the USA and Canada over shared boundary waters in 1909, and the USA and Mexico over the shared waters of the Rio Grande in 1906.

More recently, MEAs have increasingly been concerned with issues of pollution, environmental degradation, and the global commons. The UN Conference on the Human Environment, held in Stockholm in 1972, helped establish the **United Nations Environmental Program (UNEP)** and created a public awareness of the global nature of many of these problems. Many of the environmental agreements struck to date have been under the auspices of UNEP; others have been pursued on a more localized basis between neighboring countries, reflecting the regional scope of the problem.

One example of an environmental problem addressed through MEAs is the **long-range transport of pollutants**. These are pollutants that can travel significant distances before deposition and can cause air pollution, marine pollution, or water pollution. One of the best known examples of such a problem is acid rain in both Eastern North America and in Scandinavia, which is caused by industrial activity elsewhere. In the case of Eastern Canada and the USA, the **acid rain** is caused by the emission of coal-fired power plants in the US Midwest that create sulfur oxides by burning coal. Prevailing weather patterns carry the emissions several hundred miles to the east and north, where the sulfur compounds react and eventually become deposited along with precipitation, making the rain highly acidic in places where it can have a significant effect on both freshwater lakes and forest ecosystems. Because the activity takes place outside of the countries where the costs are incurred, the countries need to reach an agreement to help solve the problem. This problem has been addressed on a regional basis in both North America and in Europe, where countries have undertaken steps to identify and reduce the amount of emissions through the installation of scrubbers and other equipment, and utilizing alternative fuels.

An example of international action to address a global environmental problem is the **Montreal Protocol** (1987) which was designed to address the issue of ozone-depleting substances. Chlorofluorocarbon compounds (**CFCs**) were widely used cooling substances because they are inert, non-toxic, and inexpensive to produce. However, it has been shown that the breakdown of CFCs in the atmosphere is causing a thinning in the **ozone layer** in the earth's atmosphere. This received widespread publicity with the discovery of a "hole" in the **ozone layer** over the Antarctic in 1985. The ozone layer reduces the amount of **ultraviolet radiation** that reaches the earth's surface, and thinning of this layer increases the amount of ultraviolet radiation reaching the earth's surface with negative effects on all life on earth. The ozone layer is a classic example of a public good; all countries benefit from its presence. The Montreal Protocol specified a series of reductions in the use of CFC compounds with an eventual ban upon the use of such substances. The Protocol is widely acknowledged as being extremely successful, and has led to a sharp reduction in the production and use of CFC compounds in recent years.

Another example of international action is the **Basel Convention** (1989). It was convened to address the issue of hazardous wastes and the perceived threat from illegal shipments and improper disposal of these. One concern that has been raised is that some countries could become **pollution havens**, such that political entities would set lower environmental standards and thereby become locations where toxic industries would locate, as well as receiving toxic waste from other countries. The objectives of the agreement are to limit the transboundary shipment between countries and to create a means of defining and tracking the movement of such wastes, as well as allowing countries to refuse to accept these wastes.

Another area that has received international attention is that of **biodiversity**. People have placed increasing value upon preserving biodiversity because of the possible richness of the genetic material embodied in wild populations, as well as the inherent value given to maintaining wildlife. One of the best known agreements is the **Convention on International Trade in Endangered Species (CITES)** (1973). This agreement is designed, through the use of permits, to regulate trade in both animals and plants that are considered endangered and, if needed, to impose export bans based upon the status of the particular species (which are listed in various appendices depending upon the degree to which they are considered threatened). In general, it is felt that CITES has been relatively successful, especially in protecting certain CITES listed species such as crocodilians. More recently, the **Convention for the Protection of Biodiversity (CPB)** (1992) has tried to develop a framework to provide a mechanism by which biodiversity can be maintained in member countries. The convention envisages a mechanism by which wealthier countries can fund efforts to preserve biodiversity in less developed countries, and establishes the rights to the ownership of genetic material found within countries so that these countries can develop an economic interest in sustaining biodiversity.

The **Framework Convention on Climate Change (FCCC)**, a convention signed by over 160 countries at the 1992 Earth Summit in Rio de Janeiro, which came into force in March 1994 following its ratification by over 50 nations, is perhaps the most high-profile international agreement today. Under the agreement, Annex I countries (more **developed countries**) committed themselves to return their emissions of **greenhouse gases** to 1990 levels by the year 2000. This commitment was modified by the **Kyoto Protocol** in 1997. The ultimate objective of the FCCC is to stabilize greenhouse gas concentrations at a level such that there is no detrimental impact upon the global climate. As can be expected, this is the classic **public goods** problem – countries would benefit from the actions taken to reduce emissions regardless of where they take place. Many **developing countries** have argued that the more developed countries should take the lead in reducing emissions and that it is unfair to expect them to forgo development opportunities (through the use of inexpensive fossil fuels like coal) that were available to developed countries in their earlier stages of development. Developed countries argue among themselves how best to measure credits and debits for **greenhouse gas** emissions because different methods of calculation can make a significant difference to countries in terms of the greenhouse gas reductions they need to make.

2. TRADE AND THE ENVIRONMENT

At the same time that an increasing number of **MEAs** are being pursued, countries are also engaged in an effort to boost trade among themselves. This effort has taken two main forms: the first has been through regional trade pacts such as the **North American Free Trade Agreement (NAFTA)**, a trade agreement between Canada, the USA, and Mexico, and other agreements such as the **European Union;** and the second has been through a more general agreement called **GATT (General Agreement on Tariffs and Trade)**. GATT was formed after the Second World War, and established a framework and series of rules for trade between signatory countries through various rounds (meetings held between member countries over the past fifty years designed to further rules within a particular area). The **Uruguay Round**, which ended with the **Marrakesh Agreement** in 1994, provided the basis for the **World Trade Organization (WTO)**, an organization designed to enforce the trade rules agreed to by member countries through adjudicating trade disputes between member countries and enforcing any judgements.

These efforts to create rules to improve trade between countries have become increasingly contentious because of a perceived conflict between the wish to liberalize trade by reducing the amount of rules and government discretion over trade, and the need to take into account environmental considerations. While **Article XX** of GATT does allow for countries to make exceptions in terms of trade for goods that pose a risk to human, animal and plant life, or if conservation measures are required for the protection of exhaustible resources, it forbids such exceptions where they are used as a pretext to create trade barriers.

The best known example of a conflict between environmental and trade issues is the **Tuna–Dolphin dispute** between the USA and Mexico. The USA had passed a law prohibiting the importation of tuna from countries where the incidental kill of dolphins exceeded a certain amount (dolphins follow tuna, and certain catching methods, such as the purse seine net, can trap the dolphins along with the tuna). A US environmental group sued to enforce the law, arguing that the nets used by Mexican fishers were killing too many dolphins, and a US federal judge agreed, banning tuna imports from Mexico. Mexico brought the dispute to the GATT panel (a special body struck to hear disputes) and the GATT panel in 1991 found that the ban contravened GATT rules (although it allowed a labeling scheme that identified dolphin-friendly catching systems to stand). While the decision was never formally adopted by GATT, because the USA and Mexico reached an agreement, the case has been used as an example where trade rules have supplanted a government's ability to make environmental policy. This concern has carried over into the WTO, which has created the **Committee on Trade and the Environment (CTE)**, which is trying to recognize environmental concerns without creating a series of exceptions that make any trade rules meaningless.

At the same time, as a result of concerns raised about the ability of countries to make environmental policy given international trade rules, an increasing number of **MEAs** are also incorporating trade measures into their agreements. Examples of such MEAs include **CITES**, the **Montreal Protocol** and the **Basel Convention** (all described above). In addition, agreements such as the **Framework Convention on Climate Change** envisage market-based mechanisms as one possible way to address environmental problems.

A variety of trade measures can be used to further environmental objectives. These measures include information-gathering measures such as labeling, reporting, and the requirement for notification and prior informed consent (for receiving toxic or potentially hazardous goods). More stringent trade measures may include the requirement of either export and import permits or licenses, as well as selective import or export bans which may be between parties to an agreement as well as trade bans with countries that are not party to the agreement. These trade measures serve various purposes, but the most important objective is to make sure that all of the environmental factors are known and presumably incorporated into the exchange when trade takes place. Other objectives may include: collecting and disseminating environmental information; monitoring certain environmental factors; and the prevention of trade diversion and industrial relocation (where stricter environmental measures simply serve to redirect production or pollution to different locations rather than leading to a change in behavior and reduction in pollution).

The effectiveness of **MEA**s depends both upon the full participation of countries that are affected by, or contribute to, the environmental problem, as well as compliance with the agreement. The absence of a critical country from, or significant non-compliance with, an agreement may mean that the environmental problem only worsens. Given the interest in expanding trade found in most countries today, the ability to enter into agreements that facilitate trade can act as a strong inducement for countries that might have reservations about the costs associated with addressing environmental issues, as well as any possible diminution in government sovereignty. In addition, by setting environmental standards and policies within these different agreements, there is evidence that other countries, even if unwilling to sign formal agreements, may still adopt similar standards and policies. Finally, to the extent that concerns about environmental issues are associated with higher income levels, increased international trade may help address environmental issues through improving people's welfare.

Environmental Systems, Dynamics and Modeling: A Primer

1. ENVIRONMENTAL SYSTEMS

Our **environment** is a product of many different plants and animals, shaped by the past, the climate, the physical landscape and, of course, human activity. As **species adapt** to circumstances, they also influence their **habitats** and reshape the **environments** in which they live. **Ecosystems** (and our planet) are also subject to **random shocks** that can radically change the course of life. Indeed, the universal constant of life is change. The mass deaths that led to the extinction of the dinosaurs almost certainly contributed to **mammal speciation** and an increase in **mammal** size. Thus, if the **Alvarez hypothesis** is correct, and if a meteor had not hit the earth 65 million years ago at the end of the Cretaceous period, it is extremely unlikely that modern humans would exist and that you would be reading this book!

One of the most important processes by which life changes over time is **natural selection**. This simply means that those members of a **species** are fortunate enough to **reproduce** will bequeath their genetic material to the next generation. Members of a **species** that by chance, or physical characteristics, or behavior are less successful at reproducing, will be less favored in the **gene** pool in the subsequent generation. In this manner, life evolves over time. The process of **evolution** involves many different paths and interactions with **predators** and **prey**, and with **parasites** and **hosts** co-evolving. For example, slower deer are more likely to be caught by, say, wolves. As a result, slower deer are less likely to be successful in reproducing and thus, over time, they are eliminated and the **species** becomes faster at running. Similarly, wolves that are less able to co-operate in their hunting may be less successful at catching game and, thus, over time wolves that co-operate (assuming that co-operation is an inherited characteristic) reproduce more and dominate their **species** gene pool. **Co-evolution** can also be in the form of **symbiosis** or **mutualism**, such as with flowering plants that provide food (in the form of nectar) for **insects** that, in turn, increase the reproductive success of the plants that attract such **insects**.

The greater the period of time, all other things equal, the greater will be the effect of **evolution** on life. Nevertheless, even in very short time periods (on a geological time-scale), **species** can evolve rapidly. For example, the widespread use (or rather misuse) of **antibiotics**, particularly in the livestock industry where animals are given low doses of antibiotics to increase weight gain, has, in a short period of time, led to antibiotic-resistant **bacteria**. Rapid changes may also occur in multicellular **species**. A classic example is the peppered moth found in the United Kingdom that, prior to the **industrial revolution**, had light-colored wings and rested on **lichen** which helped camouflage them from predators, such as birds. Increased air **pollution** killed off much of the lichen on which they rested and the lighter-colored moths became conspicuous to birds, reducing their breeding success. In turn, in a few decades, the **species**, as a whole, evolved to have much darker colored wings that were less noticeable to predators in the black background of industrial Britain.

If the golden rule of ecosystems is change then evolution's maxim is **biological diversity** – the genetic variation that exists both across **species** and among individuals within **species**. Life on earth started at least 3.5 billion years ago and began with **bacteria** that still remain, in terms of their total biomass weight, the most successful life forms on our planet. By contrast, modern humans are the latest gatecrashers to a party that has been going on for almost 4 billion years! Beginning in the **Cambrian period**, some 570 million years ago, the first multicellular organisms appeared, ultimately leading to the plants and animals we see today. Almost all the life on earth (**bacteria** deep in the earth's surface may be an exception) directly or indirectly survive from the energy that reaches us from the sun. Indeed, organisms that can **photosynthesize** (or take energy from sunlight) that first evolved some 2 billion years ago, represent a fundamental turning point of life on earth. As important is the sun to the evolution of life, so has been the spin of the earth and its tilt (that gives us seasons) and the interaction of our moon that collectively help generate the energy, mixing and turbulence in the **atmosphere** and **oceans** that help sustain life.

Along the many paths of evolution, **species** evolved into other **species** such that their archaic forms passed away. In addition, many **species** became extinct in the sense that they left no descendants. Indeed, the ratio of extinct to living **species** is likely to be at least 1000 to one. In other words, if the earth has between 10~100 million **species** alive today, between 10 thousand million and 100 thousand million **species** that previously existed are now extinct. Given the genetic variation that exists within individual members of a **species**, the diversity of life on our planet since it first began is truly staggering.

If the species alive today are just 0.1 percent of all the **species** that have ever lived it begs the question, what strategies can individuals and **species** use to ensure their survival? The most obvious mechanism for survival is to produce many offspring. Indeed, the larger the number of individuals, all other things equal, the greater is the chance a **species** has of surviving into the future. Animals that use this approach are called **r-strategists** and include most **species**. An alternative strategy is to focus less on the number of offspring and instead to increase the **probability** of offspring successfully reproducing. **Species** that follow this approach, like humans, are called **k-strategists**. These strategies, however necessary, are insufficient to ensure the long-term survival of **species**. This is because any **environment** is subject to perturbations and shocks that can lead to mass deaths. For example, life on an island could be destroyed by a volcanic eruption, or many of the **species** on an entire continent could be obliterated by a meteor, or the earth's climate could suddenly become much colder leading to mass deaths (events that have all happened more than once on our planet). Unless members of a **species** can **acclimatize** and **adapt** quickly, or the species can evolve to adjust to the changed circumstances, the **species** will become **extinct**. Thus, to ensure **resilience** to local or regional shocks or events, **species** need to have **meta-populations** or to be widely distributed spatially. In other words, **species** (no matter how numerous) that are only found in geographically limited locales are at particular risk of **extinction**.

Resilience is a term that can be applied equally well to **ecosystems** as to **species**. Interestingly, some studies suggest that increased diversity of **species** (at least to a certain point) can result in a greater **biomass** from ecosystems and may foster **stability** in the sense that plant and animal communities can "bounce back" from shocks and perturbations. **Resilience** is, in part, determined by **population dynamics**. To a greater or lesser extent, species are regulated by their **population density**. Ultimately, since no population can increase indefinitely, the total population growth rate must eventually decline with **population density**. However, at low densities, increases in density may actually increase the rate of population growth. For example, at low densities some species are highly vulnerable to predation because to survive they need to aggregate or school or herd together for protection (such as with schools of sardines or herds of zebra). Moreover, the rate of growth of populations is likely to depend on the density of other **species**, such as zebras and lions or deer and wolves, in what are called **predator–prey** relationships. These linkages across species, and with **abiotic** or non-biological factors, can lead to **non-linearities** in both growth and population numbers. For example, below a **critical point** in terms of numbers, a population may simply be unsustainable. Similarly, a change in the environment beyond a **threshold** or **critical point** (such as a decline in the **pH** level of a lake) can lead to mass deaths.

The **system dynamics** of populations and their **environments** is extremely complicated and requires an understanding at the cellular level (such as **genetics**) and systems level (such as **ecology**). It also requires a comprehension of the many cycles and **feedbacks** on earth that help maintain our ecosystems, such as the **water cycle** that refreshes and recirculates water on our planet. The feedbacks in cycles are illustrated with the **carbon cycle** (shown below) that represents an interaction between **organisms** that helps determine the concentration of **carbon dioxide** in the **atmosphere**.

Simplified Model of the Carbon Cycle

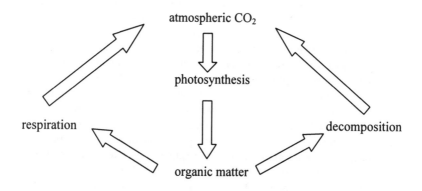

At the ecosystem level, an understanding of the **food web** is required to appreciate the **feedbacks** and links across **species** and to understand how perturbations or shocks can travel throughout an **ecosystem** or **environment**. These links can be highly complex and may even be **chaotic** such that there exists neither an **equilibrium** nor **periodicity** in how populations change over time, and the future population numbers may be highly sensitive to past values.

How life interconnects is a subject of much debate. Some writers have gone so far as to view the world as one giant interconnected organism, as in the so-called **Gaia hypothesis**. By contrast, others view **ecosystems** as merely overlapping **habitats** of different **species**. Whatever the interpretation, there is no question that how **species** interconnect plays an important role in their survival and their ability to adjust to shocks. An important step to understanding these interactions is to model the relationships and linkages among the many factors that constitute our **environment**.

2. MODELING AND DYNAMICS

Modeling is a process by which an abstract or simplification of the world around us, or events, is developed. **Models** can simply be "mental maps" or rules we develop about how the world operates. Unfortunately, the insights from models that exist in someone's head can be very difficult to impart to others. Moreover, although we are very good at expressing mental models in terms of cause and effect (I jump out of a window, I fall to the ground), we are often unable to develop mental models that reflect the **feedbacks** and interconnectedness that exist in even the simplest **systems**. Thus, to model an **ecosystem** or economic system, we need tools to help us visualize relationships and to quantify how the parts of the system affect each other. These tools include, but are not restricted to, mathematics, to help us be as explicit and as clear as possible about the nature of the **system**, and computers to help us make the many calculations required to quantify how a **system** changes over time.

Models are used for two principal purposes: **optimization** and **simulation**. **Optimization** models include an **objective function** that needs to be optimized (such as **social welfare** or species diversity) subject to a set of constraints (such as the resources available) using a set of **control** or **choice variables** (factors that can be changed, such as tax rate or the area of land in conservation areas). **Optimization** models are prescriptive and are thus part of **normative analysis** and help answer the questions of what *should* happen. **Simulation** models are used to represent a **system** or systems and, to be useful, must adequately reflect the relationships among the variables within the **system(s)**. Such models are often used for predictive purposes and help answer *what if* questions. For example, **general circulation models** in **climate modeling** help us answer the question what will be the earth's **climate** in 50 years should we continue to emit **greenhouse gases** in a **business-as-usual** scenario?

Models are often criticized as not being an adequate representation of reality. No matter how complex the model, it must (inevitably) be a simplification of the **system** being modeled. The art and science of modeling, however, is *not* to include every possible variable or relationship (which is impossible) but to include those **variables** and links that are important or significant within the system. In this sense, what is "important" or "significant" will depend on the purpose of the **model** and availability of data. For instance, a **general circulation model** used for predicting surface temperatures a century hence must include the atmospheric concentrations of **greenhouse gases** in the **atmosphere**. By contrast, a **model** used for predicting whether it will rain in a particular region in the next three days can ignore the interactions between atmospheric **greenhouse gas** concentrations and the local **weather**. Choosing what should and what should not be included in a **model** (**variables** and **feedbacks**), and whether **variables** should be treated as **exogenous** (determined outside the model) or **endogenous** (determined within the model), are questions that all modelers face. Good models are able to represent the **system** parsimoniously, or in as simple a way as possible, while still being able to effectively answer the questions for which the **model** was built.

Given that the **environment**, the **economy** and human activity are always changing, models need to be **dynamic** or inter-temporal in order to adequately reflect these systems. In other words, **variables** within the **model** should change over time. All dynamic models include **stocks** and **flows**. **Stocks** are **state variables** and may be called reservoirs, levels or inventories, depending on what **system** is being modeled. They characterize "the state of the world" such as the **biomass** of fish in a fishery model or the volume of standing trees in a forestry model or the surface temperature in a climate model. The value or level of a **stock** depends on the **initial conditions** or past values, and can change over time depending upon additions and subtractions from **flow variables**. For instance, recruitment represents a **flow** into a fishery as it increases the **stock** of fish, while harvesting of fish is a **flow** out of the fishery as it reduces the **stock**, as shown below.

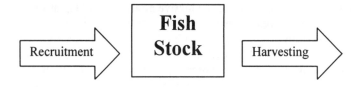

Flows represent rates of change in a **system** and could be the **birth rate** or **death rate** in a population model, or the water used and discharged by a **pulp and paper mill** in a model of water use. Although stocks are changed by inflows and outflows, **stocks** may also influence the flow variables through feedbacks. For instance, the **death rate (flow)** of a **population** is likely to be influenced by the total size of the **population** (**stock**) such that there is a feedback in the model.

Stocks that are uniquely determined in the model by flows and feedbacks are **state variables** of a system. A **model** may also include auxiliary **variables** that are functions of both **stocks** and **flows**. For instance, the recruitment into a fishery is itself influenced by a **birth rate** that is a function of both the total population (**state-determined stock**) and environmental factors (such as ocean temperature) that may be treated as **exogenous** to the **model**, as shown below.

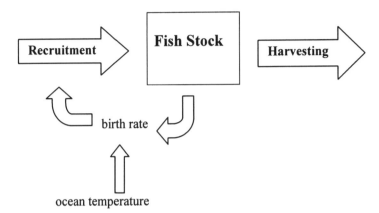

Modeling requires tests of the **robustness** and the effect of changes in a **model**, and its assumptions, in terms of results. These tests should include running or simulating the model under a variety of scenarios such as different **initial conditions, parameter** values or constants. Further, a model's predictions can sometimes (where the data is available) be compared to past values and used as a possible indicator of performance for predictions or forecasts. However, just as **correlation** between variables does not imply cause and effect, neither does past performance of a model ensure accurate predictions. In addition, a **model** should be tested to explore the effect of changes in the **feedbacks** and interrelationships of the **system**. By performing such tests and **sensitivity analysis**, researchers are better able to understand the limitations of their **models** and, thus, to comprehend better what questions the **models** can help answer and what questions they cannot resolve.

An understanding of our environment, and how to model it, is the basis of scientific research. Clearly, neither this primer nor this dictionary can do justice to the myriad of topics that encompass the many disciplines needed to understand the world around us. Readers interested in understanding more about the environment and environmental economics should, as a first step, use the references, words and appendices in this dictionary as tools in their journey of learning.

Annotated References: A Starting Point

Below is a list of annotated references that will provide a useful introduction to some of the topics covered within this dictionary. These books, in general, are not seminal works or classic texts within their disciplines, but provide a useful introduction without demanding a special expertise in the area.

The texts are grouped into eight headings, under three main areas. The first area covers the biological sciences; the second, the physical sciences; and the third, economics and modeling as it applies to environmental and natural resource issues. In addition to the texts listed below, a detailed list of references is supplied at the back of the dictionary.

1. BIOLOGY

(a) general
Campbell, N.A., J.B. Reece and L.G. Mitchell (1999), *Biology*, 5th Edn, Menlo Park: Addison-Wesley Longman.

An introduction to biology and its application in understanding how life exists on earth.

(b) cellular/genetic
Griffiths, A.J.F. (1999), *Modern Genetic Analysis*, New York: W.H. Freeman.

A more advanced text that introduces the reader to genetic theory and cellular development.

(c) evolutionary
Freeman, S. and J.C. Herron (2000), *Evolutionary Analysis*, Upper Saddle River, New Jersey: Prentice Hall.

An introductory text that examines the concepts and ideas underlying evolution.

(d) wildlife
Anderson, S.H. (1999), *Managing Our Wildlife Resources*, Upper Saddle River, New Jersey: Prentice Hall.
Focuses on the issues involved in managing wildlife in the USA, as well as providing a historical background and current trends in wildlife management.

(e) fisheries
Ross, M.R. (1997), *Fisheries Conservation and Management*, Upper Saddle River, New Jersey: Prentice Hall.

An introduction to fisheries conservation and fisheries management, including basic information on fish and their habitats.

2. BOTANY

(a) general
Salisbury, F.B. and C.W. Ross (1992), *Plant Physiology*, Belmont: Wadsworth.

An introduction to understanding how plants function.

3. FORESTRY

(a) general
Kimmins, J.P. (1997), *Forest Ecology: A Foundation for Sustainable Management*, 2nd Edn, Upper Saddle River, New Jersey: Prentice Hall.

A well written book examining the economic and biological issues involved in forest management today.

(b) silviculture
Smith, D.M., B.C. Larson, M.J. Kelty, P. Mark and S. Ashton (1997), *The Practice of Silviculture: Applied Forest Ecology*, New York: Wiley.

The techniques and application of silvicultural practices in North America.

4. ECOLOGY

(a) general
Krebs, C.M. (1994), *Ecology*, New York: HarperCollins.

An introductory text to the framework used in ecological analysis.
Smith, R.L. and T.M. Smith (2000), *Elements of Ecology*, San Francisco: Addison-Wesley Longman.

A book that examines environmental issues today from a perspective of ecological principles.

5. GEOLOGY

(a) general
Wicander, R. and J.S. Monroe (1999), *Essentials of Geology*, Pacific Grove, California: Brooks/Cole.

An introductory text to the basic principles underlying geological processes.

(b) hydrology
Fetter, C.W. (2001), *Applied Hydrogeology*, Upper Saddle River, New Jersey: Prentice Hall.

A more advanced text that outlines the basic models and science that are used in hydrology.

(c) geomorphology
Bloom, Arthur L. (2001), *Geomorphology: A Systematic Analysis of Late Cenozoic Landforms*, Upper Saddle River, New Jersey: Prentice Hall.

An introductory text that covers geological processes and the resulting landforms we see around us.

6. CLIMATE AND METEOROLOGY

(a) general

Ahrens, C.D. (2000), *Meteorology Today: An Introduction to Weather, Climate, and the Environment*, Pacific Grove, California: Brooks/Cole.

An introductory text to the basic concepts of meteorology and cycles underlying weather patterns.

Barry, R.G. and R.J. Chorley (1992), *Atmosphere, Weather & Climate*, 6th Edn, London: Routledge.

A highly informative text that covers all the principles relating to weather and climate.

(b) climate change
Houghton, J. (1997), *Global Warming: The Complete Briefing*, 2nd Edn, Cambridge: Cambridge University Press.

The definitive introduction to climate change.

7. ECONOMICS

(a) international trade
Dixit, A. and V. Norman (1980), *International Trade: Theory of International Trade: a Dual, General Equilibrium Approach*, Welwyn, Herts: J. Nisbet.

A more advanced text that uses economic theory to develop a more general model of international trade.

(b) international economics
Krugman, P. and M. Obstfeld (1997), *International Economics: Theory and Policy*, Reading, MA: Addison-Wesley Longman.

An introductory text that uses examples to examine the issues involved in international economics and international trade and finance.

(c) introductory economics
Case, K.E. and R.C. Fair (1999), *Principles of Economics*, Upper Saddle River, New Jersey: Prentice Hall.

An introductory text to microeconomics and macroeconomics.

Katz, M.L. and H.S. Rosen (1998), *Microeconomics*, 3rd Edn, Boston, MA: Irwin McGraw-Hill.

An excellent introduction to the important topics in microeconomics. The text includes a chapter on game theory and a chapter on externalities and public goods.

(d) non-market valuation
Garrod, G. and K.G. Willis (1999), *Economic Valuation of the Environment: Methods and Case Studies*, Northampton, MA: Edward Elgar.

A comprehensive overview of the main methods of pricing non-market goods and the application of the techniques in case studies that include amenity values, biodiversity and water quality.

(e) econometrics
Kennedy, P. (1998), *A Guide to Econometrics*, 4th Edn, Cambridge, MA: MIT Press.

An introduction to econometric techniques that is very readable, covers the main types of estimation techniques and addresses the most common problems encountered in empirical work.

Studenmund, A.H. (2001), *Using Econometrics: Practical Guide,* 4th Edn, Boston: Addison-Wesley Longman.

A good beginner's introduction to econometrics.

(f) natural resource economics
Hartwick, J.M. and N.Olewiler (1998), *The Economics of Natural Resource Use*, Reading, MA: Addison-Wesley Longman.

A well-written and comprehensive examination of the economics underlying the development and use of natural resources, including case studies.

Field, B.C. (2000), *Natural Resource Economics: An Introduction*, Boston, MA: McGraw-Hill.

An excellent introduction to natural resource economics that includes many examples.

(g) environmental economics
Tietenberg, T. (1994), *Environmental Economics and Policy*, New York: HarperCollins.

A very readable introduction to the economics of the environment, with many examples.

Callan, S.J. and J.M. Thomas (2000), *Environmental Economics and Management: Theory, Policy, and Applications*, 2nd Edn, Fort Worth, Texas: Dryden Press.

An excellent introduction to environmental economics and related policy issues.

(h) game theory
Bierman, H.S. and L. Fernandez (1993), *Game Theory with Economic Applications*, Reading, MA: Addison-Wesley Publishing.

A gentle introduction to game theory that includes many useful examples from economics.

(i) international development
Gillis, M., D. Perkins, M. Roemer and D. Snodgrass (1997), *Economics of Development*, 5th Edn, New York: W.W. Norton.

A useful book that examines the issues involved in economic development and includes many useful examples.

(j) cost~benefit analysis
Zerbe, R.O. Jr. and D.D. Dively (1994), *Benefit–Cost Analysis in Theory and Practice*, New York: HarperCollins.

A thorough review of cost–benefit analysis with an emphasis on applications.

8. SYSTEMS DYNAMICS

(a) environmental modeling

Ford, A. (1999), *Modeling the Environment: An Introduction to System Dynamics Modeling of Environmental Systems*, Washington DC: Island Press.

A good introduction to system modeling of the environment, with examples.

Deaton, M.L. and J.J. Winebrake (2000), *Dynamic Modeling of Environmental Systems*, New York: Springer-Verlag.

A more technical introduction to environmental modeling using system dynamics, which includes a diskette of the programs used in the book.

A

abatement. A method or process that controls or reduces discharges and emissions of pollutants.

ability-to-pay principle. The notion that publicly provided goods and services should be financed on the basis of what people are able to pay. See *willingness to pay*.

abiotic factors. Non-living components of the natural environment.

absolute advantage. The notion that an individual, firm or country is able to produce more of or a better product using identical quantities of inputs than others. See *comparative advantage (theory of)*.

absolute zero. The lowest possible temperature defined as $0°$ Kelvin or $-273°C$.

absorption. The process whereby a substance is assimilated and retained by another.

absorptive capacity. The maximum amount that can be absorbed by a substance, such as the absorptive capacity of soil for water.

abundance. A measure of the number of individuals, or groups of individuals, in a given population.

abyssal. A zone under the ocean where there is no sunlight and that is commonly defined as being deeper than 2000 meters.

acceleration principle. A hypothesis that investment varies with the rate of change in output in an economy.

access right. A property right that entitles the holder of the right access or entry to a natural resource or environmental asset, such as an entry permit to a National Park.

acclimatization. The physiological and behavioral, but not genetic, adaptations of flora and fauna due to changes in the environment. See *adaptation*.

acidic. A substance which has a low pH value. See *pH*.

acidification. The process by which a substance is rendered acidic.

acid precipitation. Precipitation in the form of snow and rain that has a pH of less than 5.0. See *acid rain*.

acid rain. Rainfall that is acidic and which has a pH level less than 5.0. The pH level of acid rain falling in eastern North America is 4.0 to 4.5. Acid rain can occur naturally but the principal cause of acid rain is the anthropogenic emissions of nitrogen oxides (NO_x) and sulfur dioxide (SO_2). Acid rain is a major problem in eastern Canada, the north-east United States, Scotland and Scandinavia.

acid shock. The biological shock that arises from the rapid acidification of water systems, such as might arise from spring runoff.

acquired immune deficiency syndrome (AIDS). A viral disease spread by contaminated blood and sexual activity. The disease has had its most devastating effects in Africa where infection rates are high and the drugs used to control the disease are unaffordable for local populations.

acre. A British measure of surface area equal to 4,840 square yards or 0.407 hectare (247.11 acres equals 1 square kilometer).

acre-foot. Amount of water needed to cover one acre of land to the depth of one foot or 0.3048 meter.

activated carbon. Carbon treated to increase its adsorption and which is used to remove impurities from water, air and solids. See *adsorption*.

activated sludge. Sludge or sewage effluent that contains a range of micro-organisms that are used to help break down solids in the secondary treatment of sewage.

active gear. Fishing or harvesting gear that must be moved through the water to be able to catch fish. An example of active gear is a trawl which is towed behind or alongside fishing vessels to catch those animals which are too large to swim through the net. See *passive gear*.

active value. See *use value*.

activity analysis. A method of analysis used to find a maximum or minimum subject to a set of linear constraints. See *linear programming*.

acute toxicity. A condition in which an organism's health is adversely affected following short-term exposure to a toxic substance.

adaptability. The extent to which a population or species can adjust to changes or shocks in the environment. See *resilience*.

adaptation. The process of genetic change of flora and fauna due to changes in the environment. See *acclimatization*.

adaptive expectations. The hypothesis that people base their expectations of the future on the basis of past outcomes. Depending upon the nature of the adaptive expectations, it can imply that people make systematic errors in their predictions. See *rational expectations*.

adaptive management. A term used to describe approaches to the management of natural resources which explicitly recognize uncertainty. Actively adaptive management requires:
1. that managers develop alternative models about how natural populations change over time;
2. that experiments (where appropriate) are used to provide more information about the relative value of the models, and
3. that these models are used to compare different policies and simulate the effects of shocks to the environment.

FURTHER READING
Walters and Hilborn (1976).

adaptive radiation. Term for the diversification and creation of different species from a common ancestor over time as each species adapts to a particular ecological niche.

ad hoc. Latin term meaning "for this purpose".

ad infinitum. Latin term meaning "without end".

ad libitum (ad lib). Latin term commonly meaning "undertaken without preparation".

ad nauseam. Latin term meaning "to the point of being sick".

adsorption.   Formation of a layer of a solid, liquid or gas on the surface of a solid without penetrating the solid. Adsorbents, such as activated carbon, are highly effective in removing undesirable pollutants from the water and air. See *activated carbon*.

ad valorem tax.   A tax calculated as a proportion of a defined value, such as a 5 percent charge on the value of timber sold.

advanced countries.   So-called advanced countries are characterized by high levels of per capita income and standard of living indicators. Advanced countries would include Japan, USA, members of the European Union and countries in the Organization for Economic Co-operation and Development. See *developing countries*.

adventure tourism.   Tourism in which tourists participate in activities that involve at least some degree of risk or danger that is perceived to be out of the ordinary. Adventure tourism usually involves participation in outdoor activities.

adverse selection.   A problem which may arise in a transaction when one party has imperfect information but the other party has full information. For example, in the second-hand car market sellers know whether their car is of high or low quality but this is difficult for prospective buyers to determine. As a result, buyers will pay the same price for high or low quality cars but this discourages sellers of high quality cars from selling. See *asymmetric information*.

FURTHER READING:
Akerlof (1970).

aerobic.   Organisms that require atmospheric oxygen to live.

aerosol.   The suspension of tiny droplets of a liquid or particles in a gas, such as air.

afforestation.   Term used to describe the conversion of land from its current use to forest use and that often occurs on marginal agricultural land.

aflatoxins.   Toxic substances that occur naturally in some molds and which are produced in conditions of relatively high humidity.

age distribution. The distribution of individuals in a population according to their age.

Agenda 21. A document that outlines policies to help achieve sustainable development and which was released at the 1992 Rio de Janeiro Earth Summit. See *United Nations Conference on Environment and Development.*

Agent Orange. A herbicide which has the toxic chemical dioxin as one of its active ingredients, a carcinogen that is suspected of causing birth mutations, skin diseases (such as chloracne) and problems with the immune system. Agent Orange was used by the American forces in the Vietnam war as a defoliant to reduce the tree and vegetative cover available to the Viet Cong. See *2,4,5-T.*

age structure models. Models of population dynamics whereby individuals are grouped according to their age.

agglomeration economies. Economic benefits and synergies that arise from the proximity of firms of the same type to each other. For example, high technology firms choose to locate in the "Silicon Valley", despite high land costs, because of site-specific agglomeration economies.

aggregate consumption. Total expenditures on consumption in an economy in a given period of time.

aggregate demand. The total expenditures within an economy in a defined period of time.

aggregate production function. A relationship that relates aggregate output to the factors of production including land, labor and capital.

aggregate supply. The total value of all goods and services produced in an economy over a defined period of time.

Agreement on Technical Barriers to Trade (TBT). A World Trade Organization (WTO) agreement that determines which non-tariff barriers to trade are acceptable and which are not.

Agreement on the Application of Sanitary and Phytosanitary Measures (SPS). An agreement that is part of the World Trade Agreement (WTO) that allows countries to restrict imports to provide protection against pest and diseases, and risks associated with chemicals, fertilizers, animal feeds, pesticides and herbicides.

agrochemicals. Chemicals used in the agricultural industry such as herbicides, pesticides and artificial fertilizers.

agroforestry. The cultivation of trees as an agricultural crop, such as the use of leaves as litter and animal feed, and the use of branches for fuel.

AIDS.   See *Acquired Immune Deficiency Syndrome.*

air.   A common name for the constituent gases in the earth's atmosphere that consists of nitrogen (78.08 percent), oxygen (21.95 percent), argon (0.93 percent) and trace gases, including greenhouse gases.

air quality control region (AQCR).   Regions in the USA used to delineate communities with similar air pollution problems, and to implement federal pollution regulations.

air quality standard.   Maximum permissible concentration of a pollutant or pollutants in the atmosphere in a defined airshed at a given period of time.

airshed.   A geographic area defined on the basis of local atmospheric conditions and used in the control of air pollution.

albedo.   A measure of the reflective ability of a body defined by the ratio of reflected radiation to incoming radiation.   Freshly fallen snow may have an albedo as high as 90 percent while black asphalt may have an albedo as low as 10 percent.

alcohol fuels.   Fuels, such as ethanol or methanol, that are produced in the fermentation of sugars or starches and that can be used to replace fossil fuels.

alcohols.   Organic compounds that are produced during fermentation and that have a wide variety of uses.

aldehydes.   Reactive compounds used in many different applications, such as in resins and glues, and which are formed by the oxidation of alcohols.

algae. A range of organisms, usually found in aquatic environments, that are capable of photosynthesis.

algal bloom. A massive proliferation of algae in water bodies due to pollution (such as phosphates), nutrification, eutrophication, and environmental factors.

algorithm. A program or set of rules that can be used to solve mathematical or computational problems.

ALH84001. Identification for a meteorite that originated on the planet Mars and landed in Antarctica some 13,000 years ago. The meteorite includes polycyclic aromatic hydrocarbons that can be produced by decaying organisms, as well as inorganic processes. Their presence, and other evidence, has been used to support the hypothesis that microbial life once existed on Mars.

FURTHER READING
Davis (1999).

alienate. The ability to transfer a good or asset to another person.

alkaline. See *base*.

allele. A specific form of a gene, that may differ across members of the same species.

FURTHER READING:
Wills (1989).

allelopathy. The process by which plants emit chemicals to prevent or retard the growth of competing plants.

Allen's rule. The so-called rule that animals in cold climates tend to have shorter and less exposed extremities than animals of related species in warmer climes. See *Bergman's rule*.

allocative efficiency. An enterprise is allocatively efficient if the ratio of the marginal products of its inputs equals the input price ratio, e.g., marginal product labor/marginal product capital = price of labor/price of capital. See *economic efficiency, marginal physical product, scale efficiency, technical efficiency*.

allopatric speciation. The evolution of new species following the geographical separation of populations.

allowable annual cut (AAC). The allowable annual harvest of timber permitted on an area of land. An AAC is often applied when the owner of the land is the state and timber companies have cutting rights on the land.

alpha diversity. The number of species in a defined area and is commonly used as a measure of biodiversity.

alpha radiation. The least dangerous form of the three main types of nuclear radiation. It consists of a stream of alpha particles released from radioactive isotopes. See *beta radiation* and *gamma radiation*.

alternative fuels. Energy derived from sources other than hydro-carbons (e.g., fuel cells, methanol).

alternative hypothesis. The contrary hypothesis to the null hypothesis in a statistical test and often denoted by H_1. For example, if the null hypothesis to be tested is that the means of two population are equal, the alternative hypothesis is that the means are not equal. If we reject the null hypothesis in a statistical test we (conditionally) accept the alternative hypothesis. See *hypothesis* and *null hypothesis*.

altruism. A term used to describe deliberate act of individuals that benefits others without the expectation of it being reciprocated.

Alvarez hypothesis. The hypothesis that the earth was hit by at least one body from outer space some 65 million year ago and that this event led to mass extinctions.

Amazonia. An area defined by the watershed of the Amazon River and including the largest area of tropical rainforest in the world.

ambient. The background level of an environmental measure, such as the air temperature or the level of particulate matter in the air.

ambient permits. Pollution permits denominated in the amount of a of pollutant that is permitted at defined receptor sites in the environment.

ambient standard. A standard that regulates the maximum permitted pollution at defined receptor sites in the environment.

amenity value. The non-market value attached to an environmental asset, such as the value of walking in a forest.

amensalism. An interaction where one organism's growth is inhibited by another. Such interaction is common among plant species.

ammonia. A gas used in the manufacture of plastics and fertilizers that when liquified at a low temperature can be used as a refrigerant. Ammonia naturally occurs in the environment as a result of microbial breakdown of proteins.

Amoco Cadiz. Name of an oil tanker that ran aground in France in 1987 and released over 220,000 tons of crude oil, which devastated the Britanny coast.

amoeba. A one-cell organism.

amortization. The provision of the repayment of a debt (such as a mortgage) over time. Amortization may also refer to the depreciation of an asset or good.

amphibian. A common term for toads, frogs and other members of the class *Amphibia*.

anabatic wind. A flow of warmer air in a valley directed upslope.

anadromous fish. Fish that live primarily in salt water but migrate to freshwater in order to spawn. Salmon is a commercially important anadromous fish.

anaerobic. An environment that contains no oxygen. Organisms that can live in anaerobic conditions are called anaerobes.

analysis of variance (ANOVA). A statistical method for analyzing the sources of variance in studies or experiments.

ancient forest. Forests characterized both by species longevity and large intervals between catastrophic disturbance. Some trees in ancient forests can attain lifespans in excess of 600 years.

angiosperm. Any flowering plant including grasses and deciduous trees.

animal unit (AU).   A measure of the amount of feed used by grazing animals set at 1,000 pounds (454 kilograms), the combined weight of one cow and one calf .

annex b countries.   Signatories to the 1997 Kyoto Protocol that collectively agreed to reduce their GHG emissions to 5 percent less than 1990 levels by 2008~12.

annex one countries.   Signatories to the Framework Convention on Climate Change (FCCC) which committed to stabilize their emissions to 1990 levels by the year 2000. See *Framework Convention on Climate Change.*

annex two countries.   Countries from the Organization of Economic Co-operation and Development (OECD) that are listed in the Framework Convention on Climate Change. Annex two countries are expected to help developing countries develop national reports on their GHG emissions and promote the use of more energy efficient technologies.

annual.   A plant that lives only one season.

annuity.   A constant annual payment. Annuities can be obtained by paying a lump sum to a financial institution in return for a constant and certain stream of annual payments.

Antarctic Treaty.   A treaty first signed in 1959 that encourages scientific research in the Antarctic in the absence of territorial and sovereignty claims by nations.

anthropocentric ethic.   Term used to describe a set of human beliefs and associated behaviors that are focused upon the well-being of human beings, to the possible exclusion of other species.

anthropogenic.   A term that indicates that an object is of human origin. For example, anthropogenic greenhouse gas emissions arise from human activity, such as the burning of fossil fuels.

anthropogenic emissions.   Term used to refer to emissions of greenhouse gases that are generated by human activity. See *greenhouse gases.*

anthropogenic fires.   Fires started by humans, often to maintain ground cover of a desired vegetation type, such as savannas.

anthropomorphic. Term used to describe the attribution of human emotions and responses to other animals. Anthropomorphism has long been criticized by biologists despite the fact that recent animal studies suggest that some life forms (such as elephants) may, indeed, experience similar emotions to human beings. For example, elephants appear to exhibit a form of grief following the death of a member of the family that has similarities to grief exhibited by human beings.

antibiotic. Chemicals used to control diseases caused by bacteria and fungi. The widespread use of antibiotics began after World War II. The over-use of antibiotics has hastened the development of antibiotic-resistant strains of bacteria.

antibody. Proteins produced by cells that can neutralize and destroy viruses and bacteria. A heightened level of antibodies due to environmental sensitivities can be deleterious to health and is associated with allergic reactions.

anti-dumping. Legislation and regulations designed to prevent foreign firms from selling goods at a price below cost.

antigen. Any substance that results in an immune response in multicelled organisms.

anti-trust. USA legislation designed to restrict the use of monopoly power and control anti-competitive behavior.

AOSIS. See *Association of Small Island States.*

APEC. See *Asia-Pacific Economic Cooperation.*

applied science. Research that seeks to use basic theories to solve practical problems.

appreciation. The increase in price or value of an asset or good over time.

appropriability. The ability of individuals to benefit from the flow of benefits from an asset or resource.

aquaculture. The managed raising and harvesting of fish and shellfish.

aquatic angiosperms. Flowering plants found in aquatic environments.

aquifer. A porous body of rock or sediments capable of storing water. Over-use of aquifers by withdrawing more groundwater than is replenished naturally is a major concern in many countries.

arable land. Land suitable for the cultivation of crops.

Aral Sea. An inland sea in central Asia located on the borders of Uzbekistan and Kazakhstan. The diversion of the waters entering the sea for irrigation purposes has reduced the water volume of the sea by a third and has contributed to salt contamination of arable land near the sea.

arbitrage. The purchase of goods or assets at one location to sell at another so as to generate a profit. See *speculation*.

archaea. A domain of microorganisms that may be some of the earliest life forms on Earth. Archaea include organisms capable of living at extreme temperatures.

Archaen era. A geological era from 3,800 to 2,500 million years ago that began with the first life on earth.

Arctic haze. Term given to an apparent haze in the winter months in the northern polar regions that is associated with air pollution in the mid latitudes.

arithmetic growth. Growth which increases by a constant amount per unit of time. For example the series 2, 4, 6, 8, ... exhibits arithmetic growth, as the next number in the series is always the current number plus 2. See *geometric growth*.

arithmetic mean. The sum of the values of all observations in a sample or population divided by the number of observations. See *geometric mean*.

aromatic hydrocarbons. Group of hydrocarbons, including benzene, so named for their strong odor. See *hydrocarbons*.

Arrow–Debreu equilibrium. A general equilibrium named after two Nobel Laureates in Economics, Kenneth Arrow and Gérard Debreu, who developed the mathematical proofs for the existence of a general market equilibrium.

artesian. An aquifer where water, because of pressure, can rise to the surface.

artificial (forest) regeneration. The establishment of new forests by artificial means, such as the planting of seedlings.

artisanal fisheries. Fisheries, often located in developing countries, where the methods of fishing are small scale and limited to coastal or inshore areas.

asbestos. A general name given to a group of materials derived from silicate minerals which are resistant to heat, and which were extensively used to provide thermal insulation and the primary material in brake pads. Its use in thermal insulation is now prohibited in a number of countries because of the risk associated with chronic lung diseases and lung cancer from repeated exposure to asbestos fibers.

ASEAN. See *Association of South East Asian Nations*.

asexual reproduction. Reproduction without fertilization.

Asian crisis. A financial crisis that began in Thailand in 1997 and subsequently affected other Southeast Asian countries and, in particular, Malaysia, South Korea, and Indonesia over the period 1997–1999. The crisis resulted in large falls in the value of the currencies and stock markets of the affected countries and led to tens of billions of dollars being granted in financial aid from the International Monetary Fund (IMF), Japan, USA, and other countries.

Asian Development Bank (ADB). Bank headquartered in the Philippines that has as its mandate to foster economic growth in the Asia and Pacific region.

Asia-Pacific Economic Cooperation (APEC). A forum formed in 1989 consisting of Pacific-rim countries including Australia, Brunei, Canada, Chile, China, Indonesia, Japan, Malaysia, Mexico, New Zealand, Papua New Guinea, the Philippines, Singapore, South Korea, Taiwan, Thailand and the USA.

aspect. The compass direction in which an area faces. Aspect is an important consideration in the planting of crops in temperate regions.

assimilation. Ability of the environment to absorb pollution or wastes.

assimilative capacity.   The maximum amount of a pollutant that can be absorbed, neutralized or made inert by the environment in a given period of time.

assimilation energy.   Measure of the energy assimilated by animals and defined as the sum of net production and respiration. See *net production* and *respiration*.

Association of Small Island States (AOSIS).   A grouping of small island nations that are likely to suffer the greatest proportional costs associated with global warming. AOSIS members have consistently argued for the need to control anthropogenic emissions of greenhouse gases.

Association of South East Asian Nations (ASEAN).   An   association formed in 1967 that now includes Brunei, Burma, Indonesia, Laos, Malaysia, Philippines, Singapore, Thailand and Vietnam. ASEAN members are committed to reducing tariffs among member countries.

asteroid.   A rock much smaller in size than a planet, found in the solar system.

asymmetrical games.   Games or strategies undertaken by individuals where the information available among the players is not the same. See *asymmetric information*.

asymmetric information.   A situation where one of the parties to a transaction does not have the same information about the transaction as another party. For example, persons seeking health insurance are likely to know more about the state of their health than an insurance company. See *moral hazard*.

FURTHER READING
Akerlof (1970).

asymptotic efficiency.   Term used to describe an estimated parameter in a statistical model that is normally distributed and with a mean value equal to the true value of the parameter.

atmosphere.   The body of gas and liquid that surrounds the earth, consist of 78.08 percent nitrogen, 21.95 percent oxygen, 0.93 percent argon and small amount of trace gases including carbon dioxide. The atmosphere is divided into four zones: the troposphere, stratosphere, mesosphere and thermosphere.

atmospheric deposition.   The deposition of nutrients and pollutants from the atmosphere, usually in the form of precipitation.

atmospheric pressure.   The total force per unit area directed by the motion of atmospheric gas molecules. Low pressure areas often lead to unstable and stormy weather while high pressure areas often provide stable weather.

attractor.   A set of points to which a dynamic system converges.

autarky.   National economic policy that precludes international trade.

autocorrelation.   Also known as serial correlation. Autocorrelation exists when errors in a model are correlated. A common measure of first-order autocorrelation (correlation between the error term and its first lag) is the Durbin–Watson statistic where a value of 2 indicates no autocorrelation.

FURTHER READING
Kennedy (1998).

autonomous.   A dynamic system where time is not an independent variable.

autoregressive process (AR).   A process by which the current value of a random variable depends on the past values of the random variable and a disturbance term. An nth order autoregressive process is defined by
$$x_t = a_1 x_{t-1} + a_2 x_{t-2} + \ldots + a_n x_{t-n} + e_t$$
where x_t is the random variable in time t and e_t is white noise. See *autocorrelation* and *white noise series*.

autotroph.   An organism capable of obtaining its nutrients directly from inorganic materials or substances.

average.   See *arithmetic mean*.

average cost.   The total costs of production divided by the total output. Economies of scale result in a downward sloping average cost curve.

average-cost pricing.   The pricing of goods sold by firms that just covers the average cost per unit of production.

average fixed cost.   The fixed costs of production divided by the total output.

average product. The total output or production divided by the amount of an input required to produce the output.

average propensity to consume. The proportion of total income spent on consumption.

average propensity to save. The proportion of total income that is saved.

average revenue. The total revenue of a firm divided by its total output.

average variable cost. The costs of production that vary with output, divided by total output.

averting behavior costs. Economic costs associated with actions or behaviors undertaken to avoid negative consequences of environmental phenomena, especially if caused by environmental change. Often cited as a proxy for the willingness to pay to avoid environmental degradation.

averting expenditures. Market costs incurred in averting the effect of a deterioration in the environment. For example, the purchase of bottled water, because of concern over the quality of drinking water, represents an averting expenditure.

AVHRR. Advanced Very High Resolution Radiometer. A sensor carried aboard NOAA meteorological satellites that provide images commonly used for vegetation mapping.

avoidance costs. The costs involved in preventing degradation of environmental quality.

B

back-end charge. A charge that must be paid for the disposal of waste.

backstop resource. A natural resource that is currently too expensive to be used but, should the price of its substitutes rise sufficiently, would become economically efficient to exploit. See *backstop technology.*

backstop technology. A technology that is available but is currently inefficient to use at existing prices. For example, the technology exists to convert coal to petroleum and was, in fact, used during World War II. However, at current prices of oil, it is uneconomic to use the technology.

backward bending supply. Used to describe circumstances under which the expected supply of a good or service falls as the price paid for the good or service increases beyond a certain point. Distinguished from normal supply curves in which the quantity supplied increases (or at least does not decrease) as the price increases.

bacteria. A group of one-celled organisms capable of multiplying by forming spores or by fission and which lack a distinct nucleus surrounded by a membrane.

bacteriaphage. A type of virus that feeds on bacteria.

bag limits. Limits imposed on hunters in terms of how many animals they are allowed to catch or kill. Bag limits are used as a control on the total level of exploitation of a population and are frequently used in recreational fisheries.

bag-tag. A system for the disposal of household waste in which only bags that are appropriately tagged are disposed. In some municipalities, households receive a fixed number of tags per week, month or year, covered in their property taxes, and additional tags can only be acquired at a much higher marginal cost.

balanced budget. A budget or forecast where a government's revenues equals its expenditures.

balanced growth. Economic growth where all sectors of the economy grow at the same rate.

balance of payments.   A method of accounting for all debit and credit transactions of a domestic country with foreign countries. Transactions include the current account, or the trade in goods and services and profits and interest earned overseas net of what was paid overseas, and the capital account, which includes the inflows and outflows of loans and investments.

balance sheet.   A financial statement which lists the assets and liabilities of an enterprise.

baleen whales.   Sub-set of the order of *cetacea* called *mysticetes* that includes the largest animal on earth, the blue whale. Baleen whales feed through bonelike plates called baleen which trap small fish and shrimp. Baleen whales are divided into three families: *balaenidae* that include right whales which swim slowly and feed by swimming back and forth through schools of their prey, *balaenopteridae* that include the blue whale and *eschrichtiidae* that includes only the Pacific gray whale which feeds along the ocean bottom. See *cetaceans*.

BANANA.   An acronym for Build Absolutely Nothing Anywhere Near Anybody, used to describe a more extreme example of public unwillingness to accept any form of development within an area because of perceived deleterious effects (which could range from increased traffic congestion due to the construction of new homes to perceived environmental degradation from a new landfill site). See *NIMBY* and *Locally unwanted land use (LULU)*.

bang-bang control.   Term which refers to how the control variable suddenly changes at a certain point in time in an optimal control problem. See *optimal control theory*.

banking.   A term for a system where polluting firms are credited for the difference between their actual and permitted emissions. The credited emissions may then be used at a later date or possibly traded, depending upon the rules of the emissions trading program.

bargaining.   The interaction between parties to a transaction.

barrel (bbl).   Standard unit of volume for oil and petroleum products that equals 159.1 liters. A ton of crude oil approximately equals 7 barrels.

barriers to entry. Factors which prevent firms from entering or competing in a market. A barrier to entry in the production of large commercial airliners is the huge capital cost required to design, build and test such aircraft.

base. Base substances have a high pH value and can neutralize acids. See *pH*.

Basel Convention on the Control of Transboundary Movement of Hazardous Wastes and their Disposal. A 1992 convention, often known as the Basel Convention, which ultimately led to an agreement to phase out the dumping or inappropriate disposal of hazardous wastes in developing countries.

basement complex. A geological formation composed of basement rocks from the Precambrian period. Also known as Archean Complex.

base period. The period in an index to which all other index values are comparable. Base periods for indexes may vary with changes in the composition of the items that make up the index, such as a change in the composition of goods used to determine a consumer price index.

basic science. Scientific research that focuses primarily upon understanding the basic principles or theories of a discipline. See *applied science.*

Batesian mimicry. An evolutionary phenomenon where a harmless organism closely resembles or mimics another species that is harmful to potential predators or is unpleasant to eat. See *Mullerian mimicry*.

Baumol effect. Idea developed by William Baumol which states that as an economy matures productivity gains will be greatest in capital-intensive sectors thus reducing the price of goods produced in these sectors relative to services which are labor intensive. As a result, public spending which includes mostly expenditures on services will, over time, account for an increasing share of the gross domestic product.

Bayesian analysis. Bayesian analysis is an approach to statistical analysis where the researcher forms a prior distribution which is then updated with additional data and Bayes theorem to generate a posterior distribution. A risk or loss function is often used to determine a parameter value from the posterior distribution. See *Bayes theorem.*

FURTHER READING
Kennedy (1998).

Bayes theorem.   Used in generating posterior distributions based on a prior distribution and available data. The theorem provides a method for updating the probability of an outcome following the acquisition of additional data or information.

FURTHER READING
Kennedy (1998).

Beaufort scale.   A system for measuring wind force developed by Sir Francis Beaufort, ranging from dead calm (0) to hurricane force (12).

Bellman's equation.   Bellman's functional recurrence equation is used in dynamic programming and defines the maximum value of a function. Its usefulness is that it provides an iterative method of solution to complex dynamic optimization problems in discrete time. See *dynamic programming*.

FURTHER READING
Chiang (1992).

bell-shaped distribution.   See *normal distribution*.

benefit-based standard.   A standard or regulation designed to lead to a desired environmental benefit, but developed with little or no consideration of the costs associated in achieving the standard. See *standard*.

benefit–cost analysis.   See *cost–benefit analysis*.

benefit–cost ratio.   The present value of benefits from an investment or project divided by the present value of the costs associated with the project. See *benefit analysis* and *present value*.

benefit stream.   The benefits that accrue over time from an investment or project. See *cost–benefit analysis*.

Benthamite social welfare function.   A social welfare function named after Jeremy Bentham, an eighteenth and nineteenth century political economist, that maximizes the aggregate utility over all individuals. See *social welfare function* and *utilitarianism*.

benthos. A term for all marine life living on or near the bottom of a lake or ocean.

benzene. A flammable liquid derived from hydrocarbons that is both toxic and a known carcinogen.

bequest value. Term used to describe a value that individuals may place on the environment that represents their consideration for its use by future generations.

Bergen declaration. A declaration signed by 84 countries in 1990. Its goal is the promotion of sustainable development.

Bergman's rule. The so-called rule that animals in colder climates tend to be larger than similar species found in warmer climates. See *Allen's rule.*

beriberi. A disease common to poor countries caused by a deficiency of thiamine. If untreated, beriberi can cause nerve damage and death.

Berlin Mandate. The decision by signatories to the FCCC to establish a protocol to deal with greenhouse gas emissions of Annex I countries beyond the year 2000. The Mandate culminated with the 1997 Kyoto Protocol. See *Framework Convention on Climate Change.*

Bernoulli distribution. Named after Jacques Bernoulli, this is the distribution of a random variable that takes on only one of two possible values per observation (with probability p for one of the values and $1 - p$ for the other) such as tossing a coin and whether it turns up heads ($p = 0.5$) or tails ($1 - p = 0.5$).

Bertrand competition. Competition between oligopolists where firms independently select the price at which they sell their product. See *Cournot competition.*

best available demonstrated control technology (BADCT). A method of determining effluent guidelines from new point sources within a source category that is based on an analysis of control technologies currently utilized in the category.

best available technology (BAT).   Term used in the regulation of pollutants. In some countries, sources of pollution are required to use the "best available" technology for controlling their emissions.

best conventional technology (BCT).   Term used in the regulation of pollutants. In some jurisdictions, sources of pollution may be required to install equipment that meets best conventional technology to achieve pollution standards rather than use the often more expensive best available technology.

best linear unbiased estimator (BLUE).   The estimator of a parameter in a linear statistical model that has the smallest variance.

best management practices (BMPs).   The best practices in an industry which are often held up as an example for others to follow.

best practicable technology (BPT).   Term used in the regulation of pollutants. In some jurisdictions, sources of pollution may be required to install equipment that meets best practicable technology to achieve pollution standards. BPT is more strict than using the best conventional technology, but is often less costly than using the best available technology.

beta.   A risk measure based on the covariance of returns of an asset and the returns from the market as a whole as a ratio of the overall variance of the market. A beta of 1 indicates that the asset, on average, moves in the same way as the market while a negative beta suggests that the asset moves in a different direction to the market.

beta diversity.   An index of the degree of change in species composition where the greater the mix the greater the index.

beta radiation.   One the three main types of nuclear radiation. It consists of a stream of beta particles released from radioactive isotopes. See *alpha radiation* and *gamma radiation*.

Bhopal.   An Indian city that was the site of a Union Carbide plant. In 1984 a leak of methyl isocyanate caused the death of thousands of individuals and affected the health of many others.

biased estimator.   A biased estimator is one where the expected value does not equal its true value.

bid rent models. Also known as random bidding models, discrete choice models in which the probability that an individual of a specified income level is the highest bidder for a specific property depends on the characteristics of the property. A different bid rent estimation is required for each income class.

biennial. A plant that germinates in one season and gives fruit in the following season.

bifurcation. A point where the qualitative behavior of a dynamic system changes.

FURTHER READING
Shone (1997).

big-bang theory. The hypothesis that the universe began 12 billion years ago in a single explosion that created time and space.

FURTHER READING
Hobson (1999).

bilaterism. An approach to international relations characterized by formal agreements between two countries. In the 1930s bilateralism focused on the granting of mutual trade privileges between two countries, which were denied to other countries. See *multilateralism*.

bilharziasis. See *schistosomiasis*.

binary choice models. See *dichotomous choice*.

binding constraint. A constraint in an optimization problem that holds as an equality at the optimum such that the value of the objective function would be greater if the constraint were relaxed. For most individuals, annual income represents a binding contraint on consumption.

BINGOs. Acronym for business non-governmental organizations.

binomial distribution. The distribution of successes (such as the number of heads when tossing a coin) in a series of trials where the probability of success is p and the probability of failure is $1 - p$.

bioabsorption.   Term applied to the process of waste removal using biological or organic material. For example, the use of moss to remove metals in a solution is a form of bioabsorption.

bioaccumulation.   The process by which persistent pollutants accumulate in the bodies of organisms. Some pollutants, such as PCBs and DDT, increase in concentration in animals higher up the food chain and can be millions of times greater than the concentrations found in animals lower in the food chain.

bioassay.   A test for the existence or concentration of a substance by using the response of an organism as an indicator.

biochemical oxygen demand (BOD).   A measure of the level of organic pollution in water where the greater the organic pollution, the larger will be the number of bacteria in the water and thus the higher the BOD. The measurement is made by calculating the oxygen utilized in a sample of water over a five-day period at a constant temperature of 20 degrees Celsius.

FURTHER READING
Tomar (1999).

biochemistry.   The study of chemical reactions in organisms.

bioconcentration.   The process by which pollutants, such as organochlorines, concentrate in the fats of animals, including human beings.

biodegradable.   A substance is biodegradable if it can be degraded or broken down over time in the natural environment.

biodiversity.   See *biological diversity*.

biodiversity hotspots.   Locations in the world with habitat that provide for a large number of different species. Most so-called hotspots exist in tropical rainforests such as in parts of Borneo, Indonesia and particular locales in Amazonia, Brazil.

biodiversity prospecting.   The cataloguing, sampling, and collecting of plant and animal species, especially with the intent of finding economically valuable uses associated with rare or previously unknown species.

biodynamic farming. A type of organic farming inspired by the writings of Rudolph Steiner. See *organic farming*.

bioeconomics. A term used to describe economic modeling that includes biological models. Many of the pioneering developments in bioeconomics came from the subdiscipline of fisheries economics.

FURTHER READING
Clark (1990).

bioethics. A view of the world which emphasizes that all life on earth is related and that human beings need to consider their partnerships with other organisms in their decision making.

biofuel. A fuel that is produced from plants or organic wastes such as wood or ethanol.

biogas. A combustible gas, consisting mostly of methane, that can be used as a fuel and is produced from the decomposition of organic matter.

biogenesis. The hypothesis that life can only be created by living creatures.

biogenic. Term used to refer to effects of biological origin.

biogeoclimactic classification system. A hierarchical system of eco-system classification that includes information on local, regional and chronological factors as well as data on climate, vegetation and other site-specific factors.

biological control. The use of biological agents, such as insects, to control pests.

biological diversity. The total variation in all life on earth or within a given area or ecosystem, typically expressed as the total number of species found within the area of interest or the genetic diversity within a species. See *extinction*.

biological growth function. The mathematical depiction of how a population behaves over time. See *population growth*.

biological legacies. Material left behind following a natural disturbance, such as dead trees after a forest fire.

biological oxygen demand. See *biochemical oxygen demand.*

biological pollution control. The use of biological systems to break down waste and effluent into environmentally benign components through natural processes.

biology. The study of life and life processes.

biomagnification. See *bioaccumulation.*

biomass. The total weight of all individuals in a given population. Biomass measures are common in species where there may be a large number of individuals, such as fisheries.

biomass dynamic models. Models that predict the total stock or biomass of natural populations. Such models have frequently been applied in fisheries and include the logistic growth model. See *logistic growth.*

biomass fuel. Organic material that can be used as a source of heat, such as wood. Since the 1970s Brazil has used sugar cane as a biomass fuel to produce ethanol, as a substitute to imported hydrocarbons.

biome. A terrestial community characterized by consistency of animal and plant species, such as the boreal forest.

biomimicry. The imitation of the designs and processes of nature to help solve problems and promote sustainable development.

bioremediation. The enhancement of the growth of micro-organisms in the environment so as to assist in the breakdown and clean-up of discharges, emissions and pollution. For example, in oil spills, the application of nitrogenous fertilizers can speed up the natural breakdown by bacteria of hydrocarbons.

Biosafety Protocol. Formally known as the Cartagena Protocol on Biosafety, this was signed by the Conference of the Parties to the Convention on Biological Diversity in May 2000. The Protocol seeks to regulate biodiversity via the regulation of the international transport of living genetically modified organisms and to assist in the protection of traditional knowledge.

biosolids. Dried solids from treated sewage that are used as a fertilizer.

biosorption. See *bioabsorption.*

biosphere. The area of land, sea and the air inhabited by living organisms.

biosynthesis. Chemical reactions produced by living organisms.

biota. Plants and animal life found in a given geographical area or time period.

biotechnology. Term used to describe the processes and technologies used to develop or enhance specific genetic traits by genetic manipulation. See *genetic engineering.*

biotic. The living community within an ecosystem.

biotic potential. Maximum possible size of a population.

biotope/ CORINE biotope. A scheme used by the European Commission to identify sites that are distinguished by their ecological diversity, richness of habitat, and relative ability to support threatened or endangered species.

birth rate. The number of births per year divided by the estimated total population, usually expressed as a rate per 1000 individuals.

bituminous coal. Most common type of coal which contains between 18 and 35 percent volatiles or substances that are emitted when the coal is heated.

bivariate data. Data where observations are available on only two variables per subject.

black economy. Part of the economy where transactions are not officially recorded so as to avoid the payment of taxes or to engage in the trade of prohibited goods. Also known as the "underground economy" or "parallel economy".

black list. A European Union list of pollutants discharged into aquatic environments and which are considered a priority for control.

blowdown.   Windthrow or uprooting of trees by the wind.

blue box.   A common term for curbside household waste recycling found in many towns and cities in North America.

blue-green algae.   See *cyanobacteria*.

board feet.   A unit used in the lumber industry to measure lumber production. One board foot is the volume of wood contained in a board 12 inches long by one inch wide by one inch high. Board feet are typically expressed in thousands and written Mbf.

bond.   A financial instrument which can be issued by companies and various levels of government, that gives the buyer a fixed amount of interest per period.

bootstrapping.   A method by which random samples are generated from an existing data set to derive confidence intervals and estimates of the parameters of interest.

FURTHER READING
Stuart and Ord (1994).

boreal forest.   A biome that includes mainly coniferous trees, characterized by long, cold winters and relatively short summers. The boreal forests of Eurasia are commonly referred to as the *Taiga*.

Boserup hypothesis.   A hypothesis developed by E. Boserup that suggests that increased population density leads to a reduction in the area of agricultural land in fallow, which in turn provides a stimulus for the development of agricultural innovations.

FURTHER READING
Boserup (1965).

botany.   The study of plants and plant life.

boundary layer (atmospheric).   Normally considered to be the part of the atmosphere most affected by the earth's surface characteristics. It can be up to one kilometer high.

bovine spongiform encephalopathy (BSE).   Commonly known as "mad cow" disease. BSE is thought to be transmitted via prions, a form of protein that has been linked to a fatal degenerative brain disease in humans called nvCJD. Transmission of BSE may have been assisted by the feeding of prion-contaminated meat and bonemeal to cattle. See *Creutzfeldt-Jakob disease.*

FURTHER READING
Fox (1998).

break-even point.   The point in production or output of a firm when its total revenue exactly equals its total cost.

breed.   An anthropogenic mating of chosen members of a species for a particular purpose. Almost all domesticated animals and plants are breeds.

breeder reactor.   A nuclear reactor capable of producing as much fuel (typically plutonium) as it uses.

Bretton Woods.   A place in New Hampshire that in 1944 was the meeting place of decision makers among the Allies who helped decide the post-World War II financial system.   The system included fixed exchange rates and the establishment of the International Monetary Fund and the World Bank.

British Thermal Unit (BTU).   The amount of heat required to raise the temperature of one pound of water one degree Fahrenheit.

broad-leaved forest.   Forest dominated by non-coniferous species such as beech or oak.

bromofluorocarbons.   See *halons.*

brownfield site.   Contaminated land in urban areas that potentially may be used for other purposes and redevelopment.

brown haze.   Atmospheric pollution distinguished by the presence of a fine brown cast to the surrounding air. The source may be microscopic particles or compounds created by photochemical reactions between various pollutants discharged into the atmosphere.

Brownian motion.   First identified by the botanist Robert Brown, this refers to a stochastic process that results in a seemingly random and erratic motion of small particles in various substances, such as liquids and gases.

FURTHER READING
Guttorp (1995).

Brundtland Commission.   The popular name for World Commission on Environment and Development and named after its Chair, Gro Harlem Brundtland. The Commission's final report called *Our Common Future* became a catalyst behind the adoption of policies to promote sustainable development.

brushing.   The use of artificial (chemical or mechanical) means to reduce competing plants in a forest so as to promote growth in desired trees.

BTU.   See *British Thermal Unit.*

btu tax.   A tax or charge based on the number of British Thermal Units in a given volume of a fuel.

bubble policy.   A sub-program of emission credit trading set up in the USA to allow some flexibility in meeting air pollution control targets. It treats multiple plants owned by the same firm as one point source in the sense that the firm can adjust its emissions among its plants to minimize its pollution abatement costs, provided that the firm's total emissions meet the defined standard.

FURTHER READING
Tietenberg (1996).

budget constraint.   A financial or revenue constraint which limits the total expenditure or spending of an individual, company or organization.

budget deficit.   The amount by which government expenditures exceed government revenues.

budget surplus.   The amount by which government revenues exceed government expenditures.

Buenos Aires plan of action.   A plan established in November 1998 by the Conference to the Parties of the Framework Convention on Climate Change which establishes a timeline and schedule of work to address the details to the principles agreed to at the Kyoto Protocol. See *Kyoto Protocol.*

buffer. Any substance that is able to prevent rapid changes in base or acidity. The ability to buffer is extremely important for animals as they can often only tolerate relatively small changes in pH within their bodies.

buffer stocks. Stocks of commodities that are held in reserve in anticipation of future events. For example, some countries maintain buffer stocks of commodities in the expectation they can sell at a higher price and some commodities, such as oil, are held in reserve for security purposes.

buffer zone. A zone of vegetation or habitat that is conserved to protect environmentally important or sensitive areas. For example, logging may be prevented on land adjacent to a national park so as to provide a buffer from potentially harmful effects that might be detrimental to animals within the park.

business-as-usual. A modeling scenario which predicts the future under the assumption that policies or actions do not fundamentally change from the present. In climate models, business-as-usual scenarios usually assume that, overall, countries fail to meet their obligations under the Framework Convention on Climate Change.

business cycle. Fluctuations in the level of economic activity where the peak is characterized by high economic growth and the trough by low growth, or possibly a recession.

butterfly effect. The notion that a small change in initial conditions, such as the flapping of the wings of a butterfly, can result in enormous effects in the future. This idea is associated with positive feedbacks and chaos theory. See *chaos* and *feedback*.

bycatch. Species that are caught or harvested incidentally when targeting other species. See *targeted species*.

C

C₃ plants. Plants, such as wheat and beans, that produce three carbon molecules with photosynthesis. See *C₄ plants*.

C₄ plants. Plants, such as corn, that produce four carbon molecules with photosynthesis and are adapted to higher temperatures and greater sunlight. The productivity of C₄ plants can be significantly enhanced with higher levels of carbon dioxide. See *C₃ plants*.

cadmium. A widely used but toxic metal that is a known carcinogen.

caesium 137. A radioactive isotope of caesium with a half-life of 30 years. Trace amounts of caesium 137 were deposited over most parts of northern Europe following the 1986 Chernobyl nuclear accident. See *Chernobyl*.

Cairns Group. A group of 18 countries, formed in Australia in 1986, organized to encourage agricultural exports.

calcrete. A layer or boundary in the soil formed by calcium carbonate.

calculus of variations. An approach to solving dynamic problems in continuous time. See *Euler's Equation*.

calorie. A measure of energy defined as the energy required to raise the temperature of one gram of water from 15.5^0 to 16.5^0 C.

Cambrian. A period on the earth 570 to 505 million years before the present.

Camelford incident. The accidental spillage of aluminum sulphate into the drinking water of the 20,000 residents in Camelford, England in 1989. The residents suffered a number of serious health complaints including symptoms of joint and muscle pain and memory loss.

campylobacter jejuni. A widespread pathogen that is the most common cause of acute infectious diarrhea in developed countries. Infections from the bacteria can be fatal and are responsible for several hundred deaths a year in the United States. The bacterium is co-adapted with poultry and thus the principal source of infection is uncooked or undercooked poultry meat.

FURTHER READING
Fox (1998).

canopy. The cover of branches and leaves in a forest. In equatorial forests the canopy may be very thick and can absorb as much as three quarters or more of the solar radiation, and over a third of the precipitation.

canopy depth. The difference in height between the lowermost and uppermost branches of trees within a stand.

capacity. The ability of a firm, in the short run, to produce the maximum level of output without any restrictions to the amount of variable inputs that can be used. Overcapacity in harvesting is a frequent problem in the management of common-pool resources. See *commons, tragedy of,* and *common-pool resources.*

Cape Domain. A region of the Cape Province of South Africa known for the high number of endemic species.

capillary action. Movement of liquids via molecular attraction, such as water through soils.

capital. Traditionally defined as human-made assets capable of producing goods and services.

capital asset pricing model (CAPM). A model of asset pricing that suggests that the returns received from an asset compensate the investor for the risks in holding that asset in a perfectly diversified portfolio.

capital consumption. The physical deterioration or reduction of capital stock through use, age or withdrawal and depreciation.

capital costs. The costs associated with plant and equipment.

capital deepening. An increase in the stock of produced capital, such as machines, relative to the stock of inputs in an economy.

capital good. A good that can be used to produce other goods, such as machines in a car factory.

capital intensive. A method of production which uses capital relatively more intensively than the other factors of production. See *factors of production.*

capital market. The so-called market where companies and governments go to seek long-term financing or loans.

capital rationing. Process whereby market imperfections or interventions restrict the amount of capital that would normally be available to borrowers.

capital stuffing. A consequence of imperfect markets and rivalry in use whereby rational firms competing for a limited share of a fixed harvest invest in gear and equipment, and other forms of capital, to achieve a competitive advantage. In aggregate, the investment in such capital reduces overall returns to the industry.

capital theory. An aspect of economic theory that refers to the optimal investment in an economy over time. Capital theoretic problems often involve the use of optimal control theory.

capital utilization rate. A measure of economic activity which refers to the rate at which capital is being used. In the long-run, a capital utilization rate of greater than 1 is not sustainable.

CAPM. See *capital asset pricing model.*

capture. The economic concept that governmental or regulatory agencies adopt the same goals and interests as the industry they regulate, and effectively lose their independence.

carbon. One of the most common elements in the environment and one which can co-form a large variety of different molecules, such as hydrocarbons and carbohydrates. Carbon is present in all organisms and can exist in a pure crystalline form such as diamonds and graphite, or in a non-crystalline form such as charcoal.

carbon-14. A radioactive isotope of carbon used in dating organic material. See *isotope.*

carbon budget. A mechanism for accounting for the net sources of carbon released into the atmosphere.

carbon cycle. A pathway along which carbon flows from the atmosphere (as carbon dioxide) through to plants in the form of carbohydrates via photosynthesis which is then metabolized by animals and then released as carbon dioxide into the air. Fossil fuels collectively represent about ten percent of the total reserves of carbon in this system.

carbon dioxide. A gas consisting of one carbon atom and two oxygen atoms (CO_2). Carbon dioxide accounts for around 0.035 percent (350 parts per million by volume) of the earth's atmosphere, which represents about a 25 percent increase over its pre-industrial level. Carbon dioxide is one of the most important greenhouse gases. See *greenhouse gases.*

carbon dioxide equivalent. For comparative purposes, greenhouse gases are sometimes converted into a carbon dioxide equivalent that represents the equivalent amount of carbon dioxide required to give the same radiative forcing. For example, the carbon dioxide equivalent of one ton of methane is 21 tons of carbon dioxide. See *greenhouse gases* and *radiative forcing.*

carbon dioxide fertilization. The effect whereby increased concentrations of carbon dioxide in the atmosphere increase the rate of growth of some plants.

carbon dioxide sinks. Depositories of carbon dioxide, such as growing forests and ocean sediments, which absorb more carbon dioxide than they release. See *carbon sequestration.*

carbon emissions. Anthropogenic emissions of carbon released into the atmosphere. The main sources of emissions include the use of fossil fuels and the clearing and burning of forests.

carbon equivalent of carbon dioxide. A measure commonly used to convert tons of carbon dioxide into tons of carbon whereby 3.7 tons of carbon dioxide equals 1 ton of carbon.

carbon leakage. The potential diversion of greenhouse gas producing industries by relocation from Annex I countries in the Framework Convention on Climate Change to non-Annex I countries that have no obligations to reduce their greenhouse gas emissions.

carbon market. A market for the purchasing and selling of credits for carbon emissions or credits for emitting greenhouse gases. See *Kyoto Protocol.*

carbon monoxide. A colorless and odorless but highly poisonous gas formed by the incomplete combustion of carbon-based fuels that is found in exhaust fumes, cigarette smoke and other sources.

carbon reservoirs. See *carbon dioxide sinks*.

carbon sequestration. The process by which the uptake of carbon dioxide in carbon sinks is increased, such as by planting trees, increasing the growth rate of trees, or by ocean fertilization. Sequestration may also involve the long-term storage of carbon or carbon dioxide in underground sites. See *carbon dioxide sinks*.

carbon sink. See *carbon dioxide sinks*.

carbon tax. A tax applied on fossil fuels based on their carbon content.

carbon tetrachloride. A toxic chemical formerly used in the dry-cleaning industry.

Carboniferous period. Also known as the coal age, this occurred from 360 to 286 million years ago.

carcinogen. A natural or synthetic substance that causes the growth of cancer cells.

carnivore. Animals, such as wolves or the big cats, the diet of which consists primarily of the flesh of other animals.

carrying capacity. Maximum number of individuals, or the biomass of a species, that can be supported by its natural environment. The term is sometimes applied to a collection of species or even the entire earth.

Cartagena Protocol on Biosafety. See *Biosafety Protocol*.

cartel. A group of producers of a commodity or a good or service whose members collaborate to set prices and/or quantity produced by each member so as to increase their returns at the expense of their consumers. The organization of petroleum exporting countries (OPEC) was a highly effective cartel in the 1970s and was able to transfer a large amount of wealth from oil-consuming countries to oil-exporting countries.

cash flow. The payments to and from a firm or organization over a given period of time. A positive (negative) cash flow exists when cash revenues exceed (are less than) cash payments.

catabolism. The breakdown at the molecular level of complex substances into simpler forms as part of an organism's metabolism.

catadromous fish. Fish that spend most of their adult life in freshwater but breed or spawn at sea, such as the American eel.

catalytic converter. A device fitted to the exhausts of vehicles that reduces emissions by helping to burn hydrocarbons and converting carbon monoxide to carbon dioxide.

catastrophe theory. The hypothesis that life on earth is subject to sudden and massive evolutionary changes due to catastrophic events.

catch per unit of effort (CPUE). A measure sometimes used in fisheries as an indicator of stock abundance, calculated by dividing the total catch by a standardized measure of fishing effort (such as days at sea). By itself, it is not a good measure of stock abundance because declines in some species are not reflected in concomitant declines in the CPUE. Further, if fishers become better at fishing over time their ability to catch fish may increase per unit of fishing effort, which may, in turn, compensate for declines in the resource.

catchability coefficient. A parameter commonly used in an aggregate harvesting function which reflects the technical efficiency of resource users. See *technical efficiency*.

catchment. An area of land which drains into a water system, such as a stream or river.

CBD. See *Convention on Biological Diversity*.

CCD. See *Convention to Combat Desertification*.

CDC. See *Center for Disease Control*.

cell fusion. The practice of combining cells from plant species that would normally not interact.

Celsius. Temperature scale named after Anders Celsius which defines the melting point of ice as 0 degrees Celsius and the boiling point of water as 100 degrees Celsius. A 1 degree change on the Celsius scale is equivalent to a 1.8 degree change in the Fahrenheit scale. See *Fahrenheit scale*.

Cenozoic. Geological era covering the time period from 66.4 million years ago to the present.

censored regression model. See *Tobit model*.

centimeter. Unit of length equal to 0.3987 inches.

Center for Disease Control (CDC). The US agency charged with promoting health by preventing and controlling disease.

center-periphery model. Model of economic development which suggests that the periphery (such as developing countries) are hindered in their development by the center (such as rich countries) through a variety of factors, including a deteriorating terms of trade. See *terms of trade*.

centralization. A process by which activities and decisions, over time, become located in a single place.

central limit theorem. The theorem states that for a random variable, with a given population mean and variance, the mean of a sample with sufficiently large enough observations will be approximately normal. Moreover, the sample mean will equal the population mean and the sample variance will equal the population variance divided by the number of observations in the sample, for a sufficiently large number of observations.

CERCLA. An acronym for the US Comprehensive Environmental Response, Compensation and Liability Act, and better known as the Superfund. The act, which came into effect in 1980, was designed to create funds to pay for the clean - up of hazardous waste sites.

CERCLIS. An acronym for the US national inventory of hazardous waste sites.

cerrado. Local term for a type of savanna found in northeastern Brazil.

certainty equivalent.   The smallest amount of money an individual would be prepared to accept for certain in an exchange for an uncertain return, such as from a lottery or gamble.

cetaceans.   Marine mammals that include whales, dolphins and porpoises and collectively represent 80 species. Toothed whales *(odontocetes)* include orca, sperm whales and all dolphins and porpoises. Baleen whales *(mysticetes)* have no teeth, instead filter feeding through plates called baleen which are lined with bristles. See *baleen whales*.

ceteris paribus.   Latin term meaning "all other things being equal". The term is frequently used in economics in partial equilibrium analysis when the effects of a change in one variable are analyzed holding other variables unchanged.

chain reaction.   A process by which a fission reaction begins in one nucleus of an atom but causes the release of at least two neutrons that, in turn, cause a fission reaction in two other atoms, and so the process continues. An uncontrolled chain reaction may result in an explosion due to a sudden and massive release of neutrons.

channelization.   The altering of stream or river flow by the construction of banks and levees and by the widening or dredging of river beds. Channelization is often done to improve navigation and transportation but can impose significant environmental costs.

chaos.   Refers to a particular behavior of dynamic systems and observed by Edward Lorenz in the 1960s. Chaos is characterized by a bounded steady state which is neither an equilibrium or periodic and where even the smallest changes in the initial conditions can lead to completely different dynamic behavior.

FURTHER READING
Gleick (1987) and Cohen and Stewart (1994).

chaotic system.   See *chaos*.

characteristic equation.   An equation the roots of which are called the characteristic roots or eigenvalues. A characteristic equation is the determinant of $A - \lambda I = 0$ where A is the matrix of coefficients, λ corresponds the eigen values and I is the identity matrix.

FURTHER READING
Grafton and Sargent (1997).

charge. See *Pigouvian tax.*

charismatic fauna. Term given to animals that have a strong appeal in soliciting support (such as baby seals) for protective measures.

chemical oxygen demand. A measure of water pollution which calculates the dissolved oxygen taken up by organic matter in a sample of water.

Chernobyl. Site, in Ukraine, of the world's worst civilian nuclear disaster. The disaster arose in April 1986 after a series of explosions in one of the reactors at the Chernobyl nuclear plant scattered radioactive debris over a large area of land.

chicken game. A game which has its origins in the actions of two individuals who are racing their cars at each other. If one of them veers away, he or she is deemed to be a "chicken" and thus loses self-esteem. If both veer away at the same time, neither gains self-esteem at the expense of the other and if neither veer away they both die. More generally, a chicken game can describe the actions of agents or countries in negotiations, such as climate change. For example, a country's reputation for defection (veering away) may reduce the likelihood of other countries from defecting (veering away) so as to avoid a potential catastrophe. See *game theory.*

chimera. Organism that has DNA from more than one species.

China syndrome. Term used for the melt-down of a nuclear reactor.

chlorinated hydrocarbons. Commonly called organochlorines, these are synthetic compounds that combine molecules of carbon, hydrogen and chlorine. Many chlorinated hydrocarbons have been used as insecticides, including DDT.

FURTHER READING
Thornton (2000).

chloroacne. A debilitating skin disease caused by contact with some chlorine-based chemicals, such as dioxins and TCDD. See *dioxins* and *TCDD.*

chlorofluorocarbons (CFCs). Compounds (such as freon) that contain chlorine, fluorine and carbon, which were widely used as refrigerants and propellants. Their use has been phased out following the discovery of the "ozone hole" over Antartica in the 1980s and the confirmation that CFCs were the principal cause of the breakdown of ozone (O_3) in the upper atmosphere at low temperatures.

chlorophyll.   Name given to the green pigment in organisms capable of photosynthesis.

choice experiments.   Experiments carried out to test assumptions about human behavior and decision making against standard economic precepts.

choice variable.   A variable for which value is solved in an optimization problem.

choke price.   1. The upper limit on the price a resource can reach before alternative resources or technologies will be used.   2. The price at which the demand for a good is just zero and for which any diminution in price will result in an at least measurable demand for that good.

cholera.   A bacterial disease that can be spread by contaminated water and food. The disease causes diarrhea and dehydration and is sometimes fatal.   The best means of prevention is adequate sanitation and housing, as well as not eating shellfish.

chromosome.   A thread of DNA found in the nuclei of plants and animals that carries the genes for inherited characteristics.

Chow test.   A statistical test named after Gregory Chow that tests for the equality of coefficients from different sample periods under the assumption of normally distributed errors. The test is frequently used to evaluate whether structural change exists over a given period of time.

FURTHER READING
Hill, Griffiths, and Judge (1998).

circular-flow economy.   1. A model of an economy which focuses on the flows of income between producers to households (through, for example, the payment of wages and salaries) and from households to producers (through, for example, consumer expenditures).   2. A model of an economic system that incorporates not only the transformation of materials into goods for human consumption, but also the generation of waste and the resulting impact on environmental systems.

CITES. See *Convention on International Trade in Endangered Species.*

civil society. A term used to describe the non-governmental organizations and individuals that contribute their time and resources for the benefit of others, as well as the good of society.

clade. Related species that share an identifiable common ancestor.

cladogram. Diagram that illustrates speciation from a common ancestor plotted against time.

class. (Plural = classes.) A classification of organisms by order. For example, *Mammalia* are a class of animals that include the order of Primates. See *order.*

class action suit. A civil law suit most commonly used in the USA whereby a lawyer or a group of lawyers sue for damages on behalf of a group of individuals, each of whom need not be named specifically in the suit.

classical economics. A term used for the body of thought developed mainly by British economists in the late eighteenth to the mid-nineteenth centuries. The best known classical economist was Adam Smith who published his *Wealth of Nations* in 1776.

clean development mechanism (CDM). Term defined in the 1997 Kyoto Protocol to assist Annex B countries achieve their targeted greenhous gas emissions reductions by reducing emissions in countries that are not obligated, under the Protocol, to undertake emissions reductions. The details of what investments and cooperation will be allowed under the CDM have not yet been agreed to by signatories to the Protocol. See *joint implementation* and *carbon dioxide sinks.*

clearcut. A forest site in which all the timber has been felled.

clearcutting. See *clear felling.*

clear felling. A forestry term used to describe a practice whereby all the trees in a given area are felled when harvesting. Clear felling, especially on steep slopes and soils prone to erosion, can impose significant environmental costs.

climactic optimum. So called optimum climate that occurred 5,000–7,000 years ago when temperatures were, on average, 2 degrees Celsius higher than they are today.

climate. Description of weather patterns over long periods of time (minimum of 30 years) as defined by the mean, range and variability of various weather measures including temperature, precipitation, sunshine hours, wind, and extreme weather events.

climate adaptation. The anthropogenic responses to mitigate the potential costs associated with climate change. Adaptation includes a multitude of actions such as the planting of more drought-resistant crops in response to a long-term decline in precipitation or the building of dykes and levees in response to a rise in sea level from a long-term rise in average surface temperatures.

climate change. Global change in the earth's climate. Much of the current debate over climate change refers to potential for global warming due to an enhanced greenhouse effect. See *greenhouse effect* and *global warming*.

climate feedback. Term that refers to the potential of one variable (or set of variables) in a climate system to affect, positively or negatively, another variable (or set of variables) which in turn may affect the original variable. For instance, a negative feedback (that reduces the rate of global warming) of increased concentration of carbon dioxide is that it can stimulate the rate of growth in some plants which, in turn, increases their uptake of carbon dioxide. A positive feedback (that increases the rate of global warming) is that initial warming may reduce snow cover in high latitudes reducing the albedo effect which, in turn, allows the earth to retain more solar energy.

climate modeling. A mathematical and systems approach to understanding and predicting the climate. The increase in computing power available to researchers since the 1980s has led to the development of complex climate models on a global scale. Global circulation models have been used to predict the potential climate changes associated with increases in the concentration of carbon dioxide in the atmosphere.

climate sensitivity. The long-term equilibrium change in mean surface temperature following a doubling of the pre-industrial concentration of carbon dioxide in the atmosphere from 280 parts per million by volume (ppmv) to 560 ppmv.

climatology. The study of climate and its variability over time.

climax community. An ecological term for the developed or mature community of species that arises after a succession of communities and which is often characterized by a greater species diversity than less mature communities. For example, an old-growth forest is a climax community that harbors a wide variety of species.

clone. The offspring of an organism that is genetically identical to its parent.

closed area. Term used in management of natural animal or plant populations whereby harvesting or hunting in an area of land or ocean is prohibited. The use of closed areas can be an important management strategy for ensuring the sustainability of animal populations and can be used, for example, to protect spawning areas.

closed economy. An economy which has very little or no trade with other countries. See *open economy*.

closed forest. Forest with at least 30 percent canopy or tree crown cover.

closed loop. A feedback control system whereby actions respond to changes in variables. See *open loop*.

closed-loop recycling. The recovery and re-use of residual products as a source of additional material for the same product, as in the use of old newspapers to make newsprint.

closed season. A period of time when harvesting or hunting is prohibited. Closed seasons are frequently used to ensure the sustainability and viability of animal populations at times when they might be highly vulnerable to capture, such as in the breeding season.

closed system. A term used to define a physical system where there is an exchange of energy with the environment but no change in mass or matter. On a grand scale, the earth and its atmosphere is a closed system. See *open system*.

club good. A good or set of goods the use of which can be restricted at little cost but is congestible such that, past a certain point, the greater the number of users the less satisfaction or utility is derived per individual from its use. The use of the facilities in a health club is a club good.

Club of Rome.   A group of people who brought their collective skills and experience together to examine the state of humanity and the earth. A famous publication of the group was the book *Limits to Growth* that predicted the medium- and long-term supply and use of natural resources.

FURTHER READING
Meadows, Meadows, Randers and Behrens (1974).

co-adaptation.   The mutual adaptation of at least two different species that is common in many predator–prey and pathogen–host interactions.

Coase Theorem.   Theorem named after Ronald Coase, a Nobel Laureate in Economics, in recognition of his insights in his paper "The Theory of Social Cost" (1960) in which he examined how the assignment of property rights can be used to overcome the problems of pollution. The Theorem is commonly interpreted as stating that the assignment of a property right can be used to internalize an externality and it does not matter, in terms of economic efficiency, which party (the polluter or persons suffering from the pollution) is assigned the right. The Theorem holds true only when people do not behave strategically, transactions costs are zero, information is perfect and the allocation of rights does not affect the marginal valuations of individuals.

coastal erosion.   Erosion of shorelines from tidal current and from storms that remove sediments.

coastal subsidence.   Subsiding shorelines due to the gradual settling of newly-deposited sediments.

Cobb-Douglas function.   A functional form, for both production and utility, in which at least two factors are combined multiplicatively (e.g., $Y = \gamma K^a L^{1-a}$). It has the characteristic that how the firm or consumer substitutes one good for another is independent of either income or production levels.

cobweb model.   A model named after the shape of the direction of the movement in the price of a good if lags exist between the supply and demand.   The dynamic model explains how prices adjust such that if supply exceeds demand, at a given price, the price will fall, which, in turn, will reduce the supply of the good in the next period leading to an increase in price. Under certain conditions, these movements in price will dampen over time and will converge to an equilibrium price where supply equals demand at the existing market price.

Codex. A set of standards on the maximum residue limits in foods devised by a committee established by the FAO and WHO.

coefficient of determination. A measure of the variation in a variable accounted for by other variables. In linear regression analysis, it is called R^2 and is the ratio of the explained sum of squares to the total sum of squares. The coefficient of determination is a frequently used measure of the goodness of fit of a model and is bounded by 0 (the model explains none of the variation in the dependent variable) and 1 (the model explains all the variation in the dependent variable).

coefficient of variation. A measure of variation of a variable defined as the standard deviation (square root of the variance) divided by arithmetic mean. The higher the value, the greater the variability of the variable.

co-evolution. See *co-adaptation*.

cogeneration. The production of both power and heat from a power source or plant. For example, a diesel engine that generates power also generates heat that can then be used for heating purposes. Cogeneration increases the energy efficiency of power plants.

cognitive ability. The capacity to fully assess all of the benefits and or costs of different choices.

cohort. Term commonly used to describe individuals of a species of a similar age or maturity. Cohorts are particularly important in fisheries management as one or two cohorts may provide most of the fish that are available for harvest.

cohort analysis. Analysis in which the focus is on a subset of a population of a particular age. See *cohort*.

cointegration. A statistical term that describes whether a linear combination of two variables is stationary such that its mean and variance are time invariant. See *stationarity* and *order of integration*.

cold war. The stand-off that existed between the United States with its allies (primarily in NATO) and the Soviet Union with its allies (primarily in the Warsaw Pact), and that resulted in the global arms race from the 1940s to the 1980s.

coliform bacteria.   Rod-shaped, anaerobic bacteria belonging to the group that includes *Escherichia coli*. Coliform bacteria usually are associated with the large intestines of vertebrates, especially mammals, and are often used as a proxy for possible human pathogens in water. See *coliform index*.

coliform index.   A measure of water pollution which counts the number of coliform bacteria in a given amount of water.

collective action.   Actions by at least two or more individuals to co-ordinate their activities. Collective action can be an important means of overcoming externalities. See *community rights*.

collective good.   See *public good*.

collective rights.   The traditional rights to a resource, area, or practice held by a community or people but which may not formally be recognized. See *community rights*.

colloid.   Very small particles dispersed in soils, such as clay.

colony.   A group of organisms of the same species that live in a mutually dependent way, such as bees in a hive.

combined sewer.   A sewer that includes stormwater and untreated sewage.

comet.   A small object made of ice and dust found in the solar system.

command and control.   An approach to pollution prevention and natural resource management that relies upon direct regulations and standards with enforcement and fines.

command economy.   An economy where much of the economic activity is directed by the state, such as was the case in the Soviet Union.

commensalism.   Interaction between at least two different organisms where one benefits from, but does not harm, the other.

commercial fisheries.   Fisheries, usually large scale, that use modern techniques and gear.

commercial forestry.   The practice of growing and harvesting trees for the purpose of producing wood products.

commercial value.   The value of a good or resource only in terms of its market value, ignoring its non-market value. Thus the commercial value of a forest is its value as timber, but this does not include its value in providing amenity and recreation.

commercial wood.   Wood that has commercial value in construction or manufacturing.

Commission for Environmental Cooperation (CEC).   A commission established under the North American Free Trade Agreement by the USA, Canada and Mexico to address environmental issues in North America, and in particular, environmental concerns due to trade liberalization.

Committee on Trade and the Environment.   A committee of the World Trade Organization (WTO) commissioned to investigate ways of reconciling measures to protect the environment and facilitating trade.

common law.   Law based on judicial precedent rather than upon enactment in legislation. Common law forms the basis of property rights in many English-speaking countries, including the United States.

common-pool resource.   A natural resource which is both rivalrous in use and from which it is difficult or costly to exclude users. If the costs of exclusion are high, or if institutional arrangements do not exist to create or enforce property rights over common-pool resources (CPRs), the exploitation of common-pool resources can result in the "tragedy of the commons" whereby the resource is overexploited in both a biological and economic sense. Despite the potential for overexploitation of CPRs, many examples exist of CPRs which have been sustainably managed, often through the use of community rights.

FURTHER READING
Ostrom (1990) and Ostrom, Gardner and Walker (1994).

common-property resource.   An often misused term that was incorrectly interpreted as natural resources where users had no property rights over the stock or the flow from the resource. Today, the term is widely accepted as defining a resource over which a community has property rights and where the community can exclude non-members from its use. See *community rights*.

FURTHER READING
Ciriacy-Wantrup and Bishop (1975).

commons.   A popular term for common-pool resources. Sometimes the term is interpreted as a common-pool resource (CPR) to which users have no property rights or a CPR over which there exists community rights. See *common-pool resource*.

commons, tragedy of.   A term popularized by Garett Hardin in his 1968 paper of the same name. The tragedy of the commons is the biological and economic over-exploitation of common-pool resources (CPRs) when there exists no property rights by users over a CPR, or flow of benefits from a CPR. See *common-pool resource*.

FURTHER READING
Hardin (1968).

communal rights.   See *community rights*.

community.   A group of animals and plants living together in a defined area.

community development quotas.   A program in the Alaska fisheries through which communities are allocated a share of the Bering Sea commercial catch.

community rights.   Property rights, defined by a community, that prevent outsiders from using natural resources and that include rules about how the members of the community can use the resources.

FURTHER READING
Ostrom (1990).

comparative advantage (theory of).   Theory of trade developed by the nineteenth-century British economist, David Ricardo. It states that countries will export those goods in which they have a relative (rather than absolute) production cost advantage.

comparative statics.   A method of analysis whereby changes to a model or a system are analyzed by assessing the changes in one or more variables following a change in another variable.

compensating surplus, variation. A measure of welfare equal to the change in income that will make an individual indifferent between some original state of the world and different state of the world, such as a different set of prices for consumption goods.

compensating wage differentials. Differences in real wages received by workers in various occupations based on their relative preferences between those occupations.

compensation principle. The idea that one can distinguish whether the move from one social state to another is an improvement, in terms of social welfare, if the benefits to those made better off by the move outweigh the losses suffered by those made worse off by the move. Determining whether this is the case requires choosing a criterion to determine how those benefits and losses are weighed against one another. See *Kaldor Potential Compensation* and *Scitovsky compensation criterion*.

compensatory growth. Natural population growth which increases (or compensates) in response to a reduction in the size of a population below its capacity. See *depensatory growth* and *logistic growth*.

competition policy. Policy, rules and legislation designed to prevent collusive or predatory business practices that are not in the long-term interest of consumers.

competitive exclusion. Biological competition between plant or animals species that results in the exclusion of one or more species from a resource or geographic area.

complement. A good or commodity that is consumed or produced with another. For example, cream is a complementary good to coffee as they are often consumed together. See *substitute*.

complements in consumption. Goods that are combined in consumption. Two goods are complements when an increase in the price of one good causes a reduced demand for both goods.

compliance. The extent to which individuals or firms conform to environmental regulations.

compliance costs. Costs of monitoring and enforcing environmental regulations.

compost.   A natural fertilizer that is a product of the aerobic breakdown of organic material.

compound interest.   See *compounding*.

compounding.   Financial term to describe a method of paying or receiving interest on both the principal (amount borrowed or loaned) and any accrued interest.

computable general equilibrium (CGE) models.   Economic models which link different sectors of the economy and that can be used to evaluate the effect of different policies. For example, CGE models have been used to assess the impact of carbon taxes on employment and income in various sectors of the economy, such as agriculture.

concave function.   A function where the area beneath the function is a convex set. See *convex function*.

concentration ratio.   Ratio which is used as an index of the market share of firms. A commonly used concentration ratio is the so-called Herfindahl–Hirschman index which sums the square of the market share of each firm and has a value of 1 if there is only one firm in the market.

condensation.   The transformation of a gas or vapour to a liquid state.

conditional probability.   The probability of an event given that another event has already occurred. The conditional probability of an event A occurring if B occurs is written as $Pr(A|B)$.

conduction.   The transfer of energy from one object to another through direct contact.

Conference of the Parties (COP).   A term for the supreme bodies that govern a number of United Nations conventions, such as the Convention to Combat Desertification (CCD) and the Framework Convention on Climate Change (FCCC). Since the ratification of the FCCC in March 1994, there have been six COP sessions, the most recent of which was held in The Hague, the Netherlands, in November 2000. The third session, or COP-3, took place in Kyoto, Japan, in December 1997 and led to the adoption of the Kyoto Protocol by the Parties. See *Framework Convention on Climate Change*.

confidence interval.   An interval defined by two numbers such that a certain probability is assigned to the true value of an estimated parameter being within the interval. Confidence intervals are calculated by using sample data from a population under the assumption that the sample is representative of the population, except for some random error. A commonly used confidence interval is a 95 percent interval where, using the sample data, there exists a 95 percent probability that the true value of parameter lies between the lower and upper bound of the interval.

congestible. Term used to describe a good or asset where beyond a certain threshold of use, further use or consumption detracts from the enjoyment of others using it. For example, at low rates of visitation to a national park an extra visitor does not affect the enjoyment of other visitors, but at high rates of visitation (such as to Yellowstone National Park in the summer), extra visitors detract from the enjoyment people receive from being there. See *club good*.

congestion. Crowding. See *congestible*.

coniferous. Plants, especially trees, that bear cones to reproduce. Many evergreen trees such as pines, cedars and firs are coniferous trees and are important species in terms of timber production in several temperate countries such as the USA, Canada, Russia and New Zealand.

conjectures. See *game theory*.

conjoint analysis. Methodologies used to model and evaluate how individuals choose bundles of goods, services and assets (including aspects of the environment) by using a set of attributes or characteristics of the choices available. The approach is based on the assumption that the choices made by individuals are undertaken by a set of decisions that identifies whether a good, service or asset meets a need, and an information-gathering exercise about the choices available to form preferences about the choices.

FURTHER READING
Louviere (1996).

connectance. A measure of the interconnectedness within a food web, defined as the number of direct species interactions divided by the total potential number of species interactions.

conservancy system. Term used to describe a sewage system that returns human excreta to the land for decomposition. The system can be effective if the level of wastes are relatively low, such that the land has sufficient assimilative capacity.

conservation. A general term for a mix of preservation, restoration, use and management of natural and human-made resources for the long-term benefit of societies.

conservation easement. A legally binding restriction on the use of a parcel of land by its owner. Frequently, conservation easements by property owners are compensated, to some extent, by reductions in property taxes.

conservation of matter. The incorrect notion that the total amount of matter or mass in a physical process must always be the same before and after the reaction or process. The idea is a useful approximation in chemical reactions and originated with the observations of the nineteenth-century scientist Antoine Lavoisier.

consistency. A statistical property of an estimator in a model. An estimator is consistent if its bias converges to zero as the number of observations tends to infinity. See *biased estimator*.

constant dollars. The value of a dollar (or other currency) in terms of what it would have been worth in real terms in a particular year or base period. For example, 1996 dollars refers to the value or price of a good in terms of how much one would have to pay for the good after accounting for the general change in the price level between 1996 and the actual period. Conversion of prices into constant dollars requires an index of changes in the general price level over time such that if a general index of prices of goods in 1997 is 3 percent higher than in 1996, the 1997 prices can be converted to 1996 constant dollars by dividing by 1.03.

constant escapement strategies. A management strategy for a natural or wild population whereby a minimum proportion of the estimated population is left unharvested so as to ensure its sustainability. Such a strategy is commonly followed in salmon fisheries where a given number of salmon are allowed to "escape", or swim upstream, so as to breed and ensure future generations of salmon.

constant harvest strategies. A management strategy for a natural or wild population whereby a fixed proportion of the estimated population is harvested over a given period of time.

constant prices. See *constant dollars*.

constant returns to scale. A situation where if all the inputs in a production process are increased by the same percentage, the output will also increase by the same percentage. See *decreasing returns to scale* and *increasing returns to scale*.

constrained optimization. Optimization (maximization or minimization) of an objective function subject to at least one constraint.

FURTHER READING
Grafton and Sargent (1997).

constructed wetlands. Wetlands purposely built for the treatment of waste water.

consumer price index (CPI). An index that tracks the expenditures, over time, associated with a defined "basket" of goods and services.

consumer species. Biological species that serve as net consumers of resources that are scarce, or subject to biological competition.

consumer surplus. The difference between the maximum that a consumer is prepared to pay for a good or service and the amount that the consumer actually pays.

consumption. The amount of goods and services used or consumed in a given period of time. Given that different goods are in different units, such as apples and oranges, consumption is measured by expenditures.

consumption efficiency. The ratio of the energy used at a given stage in the food web to energy produced at a lower level of the food web. For example, a consumption efficiency of 20 percent by herbivores implies that they ingest, or are able to utilize, one fifth of the available energy produced by plants.

consumption function. A relationship between consumption and income.

consumptive value. The value of a good or asset that is ascribed to its consumption. For example, the price of salmon sold in the supermarket represents its consumptive value, but salmon in the wild also has a non-consumptive value in terms of its value to the ecosystem and the value it may give people who like to watch salmon swim up river.

contaminant. Any substance (abiotic or biotic) that reduces the quality, quantity or longevity of another substance or material.

continental drift. See *plate tectonics*.

continental shelf. A relatively shallow and gently sloped area off a coastal land mass. The continental shelf in some places may stretch hundreds of kilometers from land (such as in north-east Australia) but, in other locations, it may end abruptly close to shore.

contingent ranking. A method for environmental valuation in which respondents rank different goods or environmental alternatives by preference. When non-market goods are ranked against market goods, the analyst can derive upper and lower bounds for the value of the non-market good.

contingent valuation. A method, usually in the form of a survey questionnaire, of eliciting values for hypothetical goods or services. The contingent valuation method (CVM) is used to estimate the value for certain environmental goods and may be the only means of estimating the passive or non-use values for goods.

continuous emission monitors. Mechanisms that track all the emissions from a particular source, rather than taking periodic samples, to calculate emissions attributable to that source.

contract curve. Based on all feasible allocations, the distribution of goods between all individuals in an exchange economy where the distribution is Pareto efficient and cannot be redistributed without making any individual worse off.

contributory value. The value of a species, resource, ecosystem or habitat in terms of the value it provides to other aspects of the environment. For example, the contributory value of a bat species could be the value it provides in pollinating flowers for commercial fruit production.

controlled waters. Rivers or streams with dams or diversionary channels to control flooding and in-stream water flow.

control variable. A variable in an optimal control problem which is under the control of the planner or programmer and which helps to determine or affects the state variable(s). For example, in an optimal control problem of a fishery, the control variable could be the level of harvest and the state variable would the fish population. See *state variable* and *optimal control theory*.

conurbation. An urban area where two or more towns or cities have grown together such that it is difficult to discern where one city ends and another begins. Where a large conurbation exists between big cities the conurbation is sometimes called a megalopolis; for example, the conurbation between Boston, New York and Washington DC is such a megalopolis and has been called Bos-Ny-Wash.

convection. The transfer of heat through a gas or liquid as the warm liquid or gas heats up, becomes less dense and rises. See *conduction*.

Convention on Biological Diversity (CBD). A 1992 convention which allows for the free trade of genetic resources while providing ways for wealthy countries to finance biodiversity conservation.

Convention on International Trade in Endangered Species (CITES). The convention was initiated in 1973 and seeks to protect endangered plant and animal species from extinction. CITES has been signed by over 100 countries and has been an important mechanism in preventing (or at least restricting) the trade in animal parts, such as ivory.

Convention to Combat Desertification (CCD). A convention that was adopted in 1994, and ratified in 1996, that recognizes societal, economic, biological and physical processes that influence desertification and focuses upon technology transfer, particularly in Africa, to address desertification problems.

convergent evolution. Evolution that leads to very different organisms that resemble each other by superficial appearance, or by behavior.

convex function. Function where the area above the function is a convex set. See *convex set* and *concave function*.

convex set. A set where if a line is drawn between any two points in the set, all the points on the line lie within the set.

co-operative game theory. Models in game theory in which the participants can make binding commitments that allow for the possibility of co-operation.

co-ordination games. An example of a common property game in which there exists two Nash equilibria: one in which players co-operate and another in which they do not.

COP. See *Conference of the Parties*.

Copenhagen Amendment. A 1992 amendment to the 1987 Montreal Protocol that accelerated the phase-out of the production of chlorofluorocarbons (CFCs) which break down stratospheric ozone. See *Montreal Protocol*.

coral. A subdivision of Colonial Cnidarians or Coelenterates characterized by the existence of symbiotic algae (*zooxanthellae*) and the tendency of many species to create habitat through calcareous, reef-forming excretions.

coram populi. Latin term for an open or public process.

core. A set of possible outcomes between individuals in a transaction where if the transaction were made all individuals would be at least as well off as if they had not made the transaction. As the number of individuals in the transaction increases, the core converges to the outcomes that would arise with perfect competition.

FURTHER READING
Cornes and Sandler (1996).

core accounts. The traditional accounts kept by countries in order to measure economic activity, including gross domestic product. See *satellite accounts*.

coriolis effect. First discovered by the French physicist Gaspard Gustave de Coriolis, and this is the apparent force experienced by an object as it moves over the surface of a rotating object. It explains why ocean currents and wind move in an apparent clockwise (anti-clockwise) direction in the northern (southern) hemispheres. This occurs because points on the equator on a rotating sphere must travel faster than points at higher latitudes. As a result, if air at the equator travelling at the same speed as the surface moves north to a higher latitude it will, initially, move faster than the air it encounters there. Thus it will appear to be moving faster eastward (in a clockwise direction) than the surface beneath it.

cornucopian. An optimistic view of the future in which resource scarcity and environmental problems do not arise and potential problems can be solved by technological progress. See *Malthusian.*

Corpus Juris Civilis. Latin term for the body of civil law.

corrective tax. See *Pigouvian tax.*

correlation. The degree to which variables move together. Negative (positive) correlation between two variables implies that a rise in one is associated with a decline (rise) in the other. See *correlation coefficient.*

correlation coefficient. A "normalized" covariance that can take on any value between -1 and $+1$ where a value of $+1(-1)$ indicates the variables are perfectly positively correlated (negatively correlated) such that they move in exactly the same (opposite) direction. See *covariance.*

correlation matrix. A symmetric matrix that provides the correlation coefficients between different variables.

corridor. An area of undisturbed forest or habitat that permits the movement of flora and fauna to ensure biological diversity and ecological integrity.

co-state variable. Sometimes called auxiliary variable, it is analogous to a Lagrangian multiplier and represents the shadow price of the dynamic constraint in an optimal control problem. See *optimal control theory.*

FURTHER READING
Chiang (1992).

cost–benefit analysis. A methodology for determining whether a project or activity generates a positive net benefit for society by evaluating all the costs and benefits over time.

FURTHER READING
Zerbe and Dively (1994).

cost-effective abatement criterion. Term for methods of pollution control that result, at equilibrium, in equal marginal costs of abatement across firms or polluters. See *marginal abatement cost.*

cost-effective policies.   Policies and practices that achieve the desired outcome at the least cost.

cost function.   A functional relationship between the level of output and the prices of inputs with the costs of production, and which may be defined as $C(y,w)$ where y is the vector of outputs and w is a vector of input prices.

FURTHER READING
Grafton and Sargent (1997).

cost minimization.   The optimization process by which costs are minimized for a given level of output. Cost minimization is a necessary condition for profit maximization.

cost shifting.   When the burden (costs) of an action (say, economic development) is shifted from the responsible actor to others who usually are separated from the actor by time, distance or scale. When the burden is shifted on to a party that cannot demand compensation or otherwise hold the actor liable, the cost shifting results in the creation of an externality.

countervailing duty.   A tariff imposed on imported goods by a country to offset an unfair competitive advantage of the exporting firm, such as a subsidy.

Cournot competiton.   Competition between oligopolists where firms independently select the amount of their product that they produce. See *Bertrand competition.*

Cournot–Nash conjectures.   A concept used in game theory. It holds that economic agents choose a strategy or set of actions which they believe will lead to the most favorable outcome based on assumptions they make about how other agents will behave.

covariance.   A measure of the joint association between two variables. If the covariance is positive (negative) the variables tend to move in the same (opposite) direction.

CPR.   See *common-pool resource.*

"cradle-to-grave" management system.   A system of managing throughput of a natural resource from initial extraction, through production and consumption, and finally to the ultimate emission of residuals to the environment.

Cretaceous. A geological period lasting from 144 million to 66.4 million years ago and encompassing both the rise and fall of the biological dominance of dinosaurs on earth. The Cretaceous period also encompassed the arrival of flowering plants.

Creutzfeldt-Jakob disease (CJD). A rare degenerative brain disease in humans. A new form of the disease nvCJD has been shown to be linked to bovine spongiform encephalopathy (BSE), or "mad cow" disease. See *bovine spongiform encephalopathy*.

criteria pollutants. Term used in the USA to define pollutants which are harmful to health and the environment.

critical depensation. Description of population growth models, where the population growth rate is negative for population levels close to zero.

critical loads. See *threshold effect*.

critical mass. The smallest amount of matter required to ensure a nuclear chain reaction. See *chain reaction*.

critical point. Threshold point beyond which further use, or increased concentrations of a substance, will significantly affect the environment See *stationary point*.

critical region. The region or set of values that will lead to the rejection of the null hypothesis in a statistical test, its size being determined by the probability of Type I error. See *Type I error* and *Type II error*.

crop rotation. Widely used practice whereby different crops are rotated around the various parcels of land on a farm so as to maintain fertility of soils and to reduce the incidence of crop diseases and pests.

cross-price elasticity. The percentage change in consumption (or production) of a good in response to a change in price of another good.

cross-sectional data. Data where observations of the variables are available for individual units of observation, but only for a given point in time. See *time-series data*.

crowding-out effect. A reduction in investment that occurs because of increased public expenditures.

crown land. Term used for public land in commonwealth countries where the head of state is the British sovereign.

crust. The outer layer of the earth that lies above the mantle.

crustal abundance. The total amount of a mineral that exists on earth.

cryptosporidium. A water-borne parasite that may survive chlorination and is becoming an increasing problem in developed countries. A recent outbreak of the pathogen in Milwaukee, Wisconsin infected over 400,000 people in 1993 and was responsible the deaths of 104 individuals.

cryosphere. The earth's ice, snow, permafrost and portion of the planet which has an average temperature of less than 0^0 C.

CTE. See *Committee on Trade and the Environment.*

cull. Timber that is not fit for sale due to defects.

culling. Management practice of removing individual animals or trees to improve the overall welfare or well-being of the population or forest. Culling is especially important for animals confined to small areas of habitat, such as national parks, where further increases in the numbers of individuals could be damaging to other species and ecosystems.

cultural control. Methods or processes for controlling natural populations by changing aspects of the environment. For example, better drainage and the removal of stagnant pools are effective cultural controls for reducing the number of mosquitoes in a city.

cumulative density function (CDF). Also known as cumulative distribution function, or simply distribution function. The CDF for a continuous random variable X that takes on the value x is
$$F(x) = Pr(X \leq x)$$
and defines the probability that X takes on a value that does not exceed x.

cumulative distribution function. See *cumulative density function.*

cumulative effect. The overall impact or effect of dosages of a substance over time.

currency depreciation. The decrease in value of a currency relative to other currencies.

current annual increment (CAI). The increase in the volume of wood from a stand of trees over the previous 12 months.

current value (dollars). Value of past returns or expected future returns in terms of today's value of money. See *constant dollars* and *discounting*.

current-value Hamiltonian. A function that represents a dynamic optimization problem in continuous time and includes the original objective function and dynamic constraints, but is free of the discount factor. See *Hamiltonian*.

cyanobacteria. Formerly classified as blue-green algae, but are now classified as a type of bacteria that produce oxygen as a byproduct of photosynthesis.

D

daisyworld. A hypothetical world described by James Lovelock to illustrate the Gaia hypothesis and how the earth can be an effective self-regulating environment for plants and animals. In the daisyworld, the temperature can be maintained between critically low and critically high levels, despite significant changes in solar radiation. Thus, by changes in the composition of black daisies (which absorb more solar radiation and do best in colder temperatures) and white daisies (which absorb less solar radiation and do best in warmer temperatures), daisyworld's temperature remains in the range that can maintain life. See *Gaia hypothesis*.

FURTHER READING
Lovelock (1990) and Lovelock (2000).

damage function. A function that relates the level of emissions or discharges to the associated environmental or social costs.

dark green technologies. Processes that are used to directly address pollution; for example, technologies used to remove surface oil on water after an oil spill. See *light green technologies*.

Darwinian evolution. Named after Charles Darwin and refers to the gradual evolution of species due to natural selection whereby individuals with characteristics that do not favor reproductive success tend to die out over time. See *natural selection*.

FURTHER READING
Darwin (1859), Gould (1989) and Stiling (1992).

Darwinian fitness. Named after Charles Darwin and also known as adaptive value. It refers to the relative ability of an individual, with a given genotype, to pass on its genes to future generations. See *Darwinian evolution*.

data fouling. Process by which unrepresentative or misleading data is used in decision making because respondents deliberately supply false information.

data mining.   An approach to estimation of statistical models by which a large number of data are used or "mined" to ensure that a given model achieves an acceptable goodness of fit.

DDT.   Acronym for dichloro-diphenyl-trichloroethane. DDT is an insecticide that was once widely used, especially for the control of malarial mosquitoes, and which is very persistent within the environment. DDT accumulates in the fatty tissues of animals. The highest concentrations are often found in animals at the top of the food chain. Its persistence and ability to accumulate in animals, causing deleterious health effects, has led to its ban in a number of countries, including the USA and the European Union, although it is still used in some developing countries.

deadweight loss.   Net economic loss to society from a policy or action that alters a competitive outcome.

death rate.   The number of deaths per year divided by the estimated total population, usually expressed as a rate per 1000 individuals.

debt-for-nature swaps.   Agreements whereby countries or conservation groups buy some of the existing debt owed by a developing country which, in return, agrees to set aside a given area of land for conservation purposes.

decentralization.   Process by which activities and decisions formerly made at one, or a few locations, become more widely distributed over a greater number of locations.

decentralized outcome.   An outcome, such as a market equilibrium, which arises from many individuals making their decisions independently.

decibel (dB).   A logarithmic measure of sound intensity. The sound of traffic is about 80 dB while the rustling of leaves may be as little as 5 dB or less.

deciduous plants.   Plants and trees that at some period during the year lose their leaves.

decision rule.   A rule whereby an event triggers a decision or response. For example, a farmer may have a decision rule to plant tomato seeds on the first full moon in May.

decision tree. A representation of the subsequent decisions that can be made and the possible outcomes of each possible decision.

declining block pricing. A pricing mechanism where increments of a good or service, such as electical power, are priced in blocks, and prices decline for addional increments as the total quantity purchased increases.

decomposition. The breakdown of organic material by microbial activity.

decreasing returns to scale. A situation where, if all the inputs in a production process are increased by the same percentage, the output will increase by a smaller percentage. See *constant returns to scale* and *increasing returns to scale*.

deduction. See *deductive method*.

deductive method. A method of reasoning by which logic and abstract theory are used to explain or predict actual phenomena. See *inductive method*.

deep ecology. A philosophical view of life on our planet that places no special or unique importance on human beings and has an holistic or even spiritual view of living systems. The term originated in the 1970s with the work of Arne Naess. See *shallow ecologists*.

FURTHER READING
Devall and Sessions (1985).

de facto. Latin term meaning "existing in fact" or in practice.

de facto property rights. Property rights that have no legal recognition, but may nevertheless exist in practice. See *property rights* and *de jure property rights*.

defaunation. The removal of animal species from a geographic area.

defensive environmental costs. Expenditures associated with overcoming the potential harmful effects of a deterioration in the environment. For example, if water quality declines, the associated increased expenditures on bottled or filtered water are defensive environmental costs.

deficit. See *economic deficit*.

deficit forest.   Forest where the current annual cut exceeds the long-run sustainable yield. See *surplus forest.*

deflation.   A general fall in prices of goods and services in an economy which may be measured by a reduction in the consumer price index.

defoliant.   A herbicide that destroys the leaves of plants and trees. See *Agent Orange.*

defoliation.   The loss of leaves, usually by chemical or incendiary means. Agent Orange was a chemical used during the Vietnam war by the United States military to cause the defoliation of tropical forests in order to deprive opposing forces of cover.

deforestation.   The transformation of land use from growing trees to other uses, such as for growing crops or to pasture land. Much of the world's deforestation is occurring in tropical and sub-tropical regions.

FURTHER READING
World Commission on Forests and Sustainable Development (1999).

degradation.   1. The diminution in value or output. Environmental degradation is the state in which the environment has been altered in some way to make it less productive or valuable from any perspective (e.g., economic, ecological, or spiritual). 2. The geological weathering of land or the breakdown of an organic compound.

degrees of freedom.   A statistical term that refers to the number of independent units of information (usually number of observations) that are left after estimating the parameters of a statistical model.

de jure.   Latin term for existing or having a basis in law.

de jure property rights.   Property rights that have legal recognition. See *property rights* and *de facto property rights*.

delta.   Deposit of sediments at the mouth of a river or stream.

demand.   The willingness or desire for a good or service at a given price. See *demand function.*

demand curve.   See *demand function.*

demand function. A relationship between the amount people are willing to spend on a good or service and the price of the good, holding everything else constant.

demand management programs. Schemes that try to reduce or redistribute demand for a good through incentives, such as higher pricing for power supplied during peak times, or rebates offered for models of refrigerators that have a lower energy consumption.

demand site management. Term used to describe the efforts, other than by price, of utilities and governments to control the amount of energy used in businesses and households. These efforts may include a variety of campaigns, such as encouraging people to turn off the lights when a room is vacated.

de minimis risk. A trivial or minimal risk associated with an activity.

demographic transition. The notion that with improvements in the standards of living and health care, death rates drop before an appreciable fall in birth rates thus increasing the rate of population increase. Eventually, economic development leads to increased job opportunities for persons with specialized skills and training, and especially for women, and the birth rate falls until the population growth stabilizes.

demography. The study of change, over time, in the structure and size of human populations.

dengue fever. A viral disease that can sometimes be fatal. It is transmitted by mosquitoes that are endemic in most tropical countries. Control is through the spraying of pesticides to reduce mosquito populations, as well as improvements in urban infrastructure to reduce areas of surface water that facilitate the breeding of mosquitoes.

denitrification. The breakdown of nitrogen compounds by micro-organisms.

density. Mass per volume and which is usually defined as grams per cubic centimetre.

density-dependent growth. Growth that is regulated by the density of a population. For example, as the population increases, the rate of mortality and morbidity may also rise, which reduces the growth in the population. In some species, such as sardines, increased density at low levels may actually increase the growth in the population as it enables fish to school together and can be an effective strategy against some predators.

density independence.   The notion that, over a certain range of population density, the growth of a population is independent of its density.

deoxyribonucleic acid (DNA).   An organic molecule consisting of two strands that form a double helix and which contains genetic information.

FURTHER READING
Frank-Kamenetskii (1997).

dependent variable.   The variable in a model or optimization problem which depends on (i.e; is a function of) the other variables. In regression analysis, the dependent variable is found on the left-hand side of the regression equation.

depensation.   In population growth models, the case in which the proportional population growth rate is an increasing function of population size.

depensatory growth.   Natural population growth which declines in response to a reduction in the size of a population below a critical level. See *compensatory growth* and *logistic growth*.

depletable resources.   Those resources that cannot be reasonably expected to be replenished by natural processes.

depleted.   The point at which extraction of a non-renewable resource from a deposit ceases. If it is the case that there is no mineral deposit left, then the mine is physically depleted. If it is uneconomic to extract any more, then the mine has been economically depleted although there may still be some mineral deposit left.

depletion costs.   The reduction in value of a non-renewable resource that results from a fall in the total quantity available due to extraction.

depletion indices.   Measures of how quickly a non-renewable resource is being used; typically measured in terms of the years remaining of known reserves based upon either current or projected rates of consumption.

deposit-refund system. An increasingly common method of encouraging the proper disposal of waste whereby the price of a good includes a deposit that is refundable, provided that the residual of the good in question is returned to a center for appropriate disposal or for recycling.

depreciation. The physical loss and/or financial loss over time associated with the use of an asset.

derby fishery. A fishery where the total catch is fixed and fishers compete for a share of the harvest. A derby fishery is characterized by "the race to fish" that dissipates the potential rents. See *common-pool resource*.

deregulation. Term used to describe the removal or reduction in the rules governing individual or firm behavior.

derived demand. A demand for a good or service that is derived from the demand for another good or service. For example, an increase in the demand for paper products may lead to an increase in the derived demand for tropical timber.

derived savanna. Savanna vegetation created by the anthropogenic alteration of another type of vegetation (e.g., forest). Most commonly used to describe human-maintained savanna in Sub-Saharan Africa.

DES. See *diethylstilbestrol*.

desalinization. A process for removing salt from water or soil. Desalinization of soil can involve better drainage and the "flushing" of soil with water with low levels of sodium. Desalinization of sea-water is an important source of drinking water in some countries, such as Saudi Arabia.

desert. An arid region characterized by little rainfall.

desertification. Process by which the moisture in the soil over an area of land and vegetative cover declines over time. Desertification can be caused by both natural (e.g., climate change) and anthropogenic factors (e.g., deforestation).

detergents. Synthetic cleaning agents used for a variety of purposes that can sometimes lead to environmental problems. In the past, phosphates were included in detergents and contributed to eutrophication of lakes.

determinant.   The sum of the product of the elements of any column or row in a square matrix multiplied by their cofactors. The determinant of a square matrix A is denoted by |A|.

deterministic.   Term used to describe a model or process where all future values can be explained or "determined" by the model or the process itself such that there is no uncertainty about the future.

detrending.   Any method that tries to remove a trend from time-series data. A common procedure is to regress the variable of interest against a constant term and regressors involving a measure of time. The residuals from the regression are "detrended" and can then be used to test hypotheses about fluctuations around the trend.

detritus.   Fragments composed of partially decomposed organisms.

devaluation.   Where countries fix the value of their currency relative to other currencies, a reduction in the value of the domestic currency relative to a foreign currency is called a devaluation.

developing countries.   Name used to refer to countries with low per capita incomes and which suffer from relatively high rates of morbidity and mortality, such as Bangladesh or Bolivia.

FURTHER READING
Ray (1998).

developing world.   See *developing countries*.

development.   A process by which individuals, communities and societies become empowered better to achieve their human potential. For example, improvements in education and health care and increases in real disposable income are often considered as factors that contribute to development.

development economics.   Study of the process of economic development on both a macro and microeconomic perspective and with a focus on developing countries. See *developing countries*.

FURTHER READING
Ray (1998).

development rights.   The right to exploit or develop a resource or area.

Devonian.   Geological period from 408 to 360 million years ago.

dewatering.   Removal of naturally occuring water from sediments by drainage.

diadromous.   Animals that migrate between salt and fresh water, such as some species of eels.

diameter.   Measurement of the size of the tree trunk. Sometimes called d.b.h. for diameter-at-breast-height.

diamond water paradox.   The observation by classical economists that although water was essential to human life and diamonds were not, the price of water was essentially zero while the price of diamonds was high. This discrepancy between the apparent value and price of the two goods was explained by the fact that there were abundant supplies of water while diamonds were in short supply.

diatom.   Type of algae that exist in aquatic and marine environments.

dichotomous choice.   Survey questions often used in contingent valuation studies in which the respondent is asked to choose one of two choices. For example, the question "Would you be willing to pay $10 to ensure the preservation of community wetlands?" requires the answer yes or no.

dieback.   A widespread death of trees and plants. Diebacks may be caused by pollutants, such as ground-level ozone and acid rain, or by natural factors.

diesel exhaust particulate matter.   Particulates made up of chains of carbon matter containing a variety of organic compounds, some of which are carcinogenic.

diethylstilbestrol (DES).   A synthetic chemical that acts like the hormone estrogen and which for many years was given to pregnant women to "normalize" their pregnancies. Recent evidence has linked DES to physical abnormalities and other health problems.

FURTHER READING
Colborn, Dumanoski and Myers (1997).

difference equation.   A representation of the changes in variables in discrete time. An example of a difference equation is

$$x_t = a + by_{t-1}$$

where a and b are parameters, x_t is the value of the variable x at time period t, and y_{t-1} is the value of the variable y at time period t - 1. The solution must be consistent with the difference equation and contain no lagged terms.

FURTHER READING
Grafton and Sargent (1997).

differencing.   A method used to eliminate trends from time-series data whereby an observation of a variable in period t is subtracted from an observation of the variable in period t+n. The degree of differencing is determined by n, where if n=1 the resulting observations are called first differences.

differential equation.   A representation of the rates of changes in variables in continuous time. An example of a differential equation is

$$dx/dt = a + by$$

where the solution must be consistent with the differential equation and contain no derivative terms.

FURTHER READING
Grafton and Sargent (1997).

differential rent.   See *Ricardian rent*.

differentiated regulation.   Standards or requirements that are not applied uniformly and may have the effect of favoring different types of technology or products (such as emission standards placed upon newer model cars that are absent for older cars).

diffusion.   1. Movement of material from areas with higher concentrations to areas with lower concentrations. 2. The adoption of ideas or practices developed elsewhere and then applied in new environments or circumstances.

dilution approach.   The method of managing pollution by not treating the waste emitted but rather diluting it as it is discharged through various means (such as discharging raw sewage directly into the sea).

diminishing marginal product.   See *diminishing returns*.

diminishing returns. A term used to describe the eventual decline in the marginal product of a variable input as more of the input is used while keeping other inputs unchanged. See *marginal physical product*.

dimorphism. Term for the existence of two distinct types of organisms within a population of a species. The most common form of dimorphism is sexual dimorphism where there exist physiological differences between males and females.

dioxins. Many dioxin compounds exist and are formed in the production of organochlorine compounds (such as herbicides) and in the manufacture of chlorine-bleached wood-pulp. Some dioxins, such as tetrachlorodibenzo-p-dioxin (TCDD), are extremely toxic and are endocrine disrupters.

direct taxes. Taxes imposed on income or wealth. See *indirect taxes*.

direct valuation. The direct measurement of changes in environmental quality attributable to different policies or actions, and their impact on individual's well-being in financial terms.

discarding. Term used in fisheries to describe the actions of fishers who dump unwanted fish overboard at sea. Discarding can be a potentially serious problem as the discards are not reported in the catches and the dumped fish often have a very high mortality rate.

discount factor. A numeric value that serves to convert a future value to a present value. See *discount rate*.

discounting. Method by which future costs and benefits are converted into current dollars. For example, if the discount rate is rpercent per year then a payment of $A after n years is, worth $A/(1+r)^n$ in current dollars, after discounting.

discount rate. In consumption analysis, a factor by which future welfare or utility is multiplied to indicate that future consumption is less valuable in the present than current consumption. See *time preference*.

discovery cost. The cost of finding new mineral deposits.

discrete choice. Models of consumer choice in which the good or alternative chosen by the consumer, is available only in discrete (e.g., integer) units.

disembodied technical change.   Technical change that arises from the development of new techniques and methods of production rather than from the use of new machines and capital goods. See *technological change* and e*mbodied technological change.*

disequilibrium.   A failure to reach a steady state.

dissolved oxygen (DO).   The amount of oxygen dissolved in water, usually specified by parts per million. Low levels of DO are deleterious to fish as well as other aquatic life and may indicate the presence of anthropogenic pollutants.

Distance-related distortions.   A failure to internalize the costs of an action or behavior due to a distance over time, space, or scale between the agent and the environmental impact.

distributional effects.   See *distributional impacts.*

distributional impacts.   The economic costs imposed upon, and benefits received by, different individuals because of policy decisions (e.g., a program to provide a price floor for wheat may benefit farmers but harm consumers through higher prices).

disturbance term.   See *residual.*

diurnal temperature range.   The difference between the maximum and minimum temperature in a 24-hour period.

diurnal variation.   The difference in sunlight, and hence energy available to organisms, and its impact on natural processes during the course of a day.

divide.   The boundary between two drainage basins.

divisibility.   A property right characteristic that represents the ability to divide a resource or asset into sub-units, such as denominating pollution permits into tons.

division of labor.   Term used to describe how, in general, individuals specializing in particular activities can become very good at that activity and thus can increase their productivity over time. Provided that people can exchange their goods and services with others at low cost, the division of labor can benefit everybody.

DNA. See *deoxyribonucleic acid.*

Dobson unit. Named after Gordon Dobson, a Dobson unit is a measure of the level of ozone concentration in the atmosphere.

domesticated. Animals and plants that have been bred for human use. Almost all the plants and animals used in agriculture are domesticated species.

dominant groups. Species that have the most influence within an ecosystem.

dose–response relationship. The measurable effect of different levels of pollution on some aspect of the environment, including people's health.

double dividend. Term used to describe the potential benefits from increasing "green taxes" on pollution or emissions with a corresponding reduction in other taxes, especially direct taxes, such that the net effect is revenue neutral. The so-called double dividend is an improvement in environmental quality and an increase in output as workers have a greater incentive to work more with reduced direct taxes.

dracunculiasis. A parasitic disease, also known as the Guinea Worm disease, found in sub-Saharan Africa and the Indian sub-continent.

drainage basin. The area of land from which all surface water flows into a single river and/or its tributaries.

drawdown. Term used to describe the reduction in the annual allowable cut of forest to its long-term sustainable level. Also known as falldown.

dredging. The removal of sediment from a water body to deepen channels or harbors.

drift net fishing. A method of fishing in which large nets attached to buoys are left in the sea for a period of time and then brought on board the fishing vessel. The practice is particularly destructive as all fish that are larger than the mesh size are caught in the net. Large-scale drift nets are banned by the United Nations.

dryland management.   An umbrella term for the management of arid and semi-arid lands, primarily for agriculture.

dummy variables.   Variables that take on value of 1 or 0 for categorical variables such as male (dummy variable = 1) and female (dummy variable = 0).

dumping.   1. The sale of goods in a market at below the cost of production. 2. The improper disposal of wastes.

duopoly.   Term given to where there exist only two sellers of a good or service in a market.

durable goods.   Goods, such as refrigerators, that provide a service over an extended period of time.

duration.   A characteristic of property rights that represents the expected longevity of the asset, or the expected time a user may be allowed to use the asset.

Durbin–Watson statistic.   A statistical test used to determine whether the residuals in a multiple regression are correlated. See *autocorrelation*.

FURTHER READING
Ramanathan (1992).

dustbowl.   A large area of land in the mid-west and south-west of the United States where air-borne soil erosion due to drought and poor agricultural practices became a huge environmental problem in the 1930s.

dynamic control.   A system in which the control variable can respond to changes in the state of the system.

dynamic efficiency.   A mode of production where firms continuously improve their practices over time. Methods of pollution control are dynamically efficient if they encourage firms to reduce emissions over time.

dynamic equilibrium.   See *steady-state*.

dynamic game.   A repeated game in which a player's strategy can span the multiple iterations of the game.

dynamic model. A model whose variables change with respect to time.

dynamic programming. A recursive method of optimization for problems defined in discrete units of time. See *Bellman's equation*.

FURTHER READING
Chiang (1992).

dynamic stability. The ability of a system or process to return to an equilibrium following a disruption or shock to the system.

E

E-4 scheme. Coined by Maurice Strong as the four variables that best describe sustainable development programs: equity, economics, ecological integrity, and empowerment.

Earth sciences. See *geology*.

Earth Summit. See *United Nations Conference on Environment and Development.*

earthworm biotechnology. The treatment of livestock effluent by first preconditioning the slurry in tanks and then applying the same to land already enriched with earthworms. Earthworm technology is thought to reduce odorous emissions from effluents and to speed the decomposition of the slurry.

easement. A legal right between a landowner and another to use the land for a particular purpose, such as the passage of telephone lines. See *conservation easement.*

Ebola. A viral disease named after a place in Zaire where it was first discovered in 1976. The disease is extremely infectious and often fatal. The host of the disease is most likely endemic to the equatorial forests of central Africa, and thus increased human activity in the forest, such as deforestation, may increase the number of outbreaks.

ecofeminism. A school of thought that places the exploitation of women by men as the fundamental explanation for the misuse and abuse of the environment. See *social ecology.*

ecolabels. The International Organization for Standardization (ISO) has established three standards for labeling "environmentally friendly" products. Type I labels are for products independently monitored and environmentally preferred throughout their life cycle, type II labels are for products where the manufacturers make claims about whether the product is environmentally friendly but without accepted criteria for their use, and type III labels provide factual information about the product but provide no judgement as to whether they are environmentally friendly or not.

E. coli. See *Escherichia coli.*

ecological balance. The condition in which a community of organisms or an ecosystem is in dynamic equilibrium, within which genetic species, and ecosystem diversity remain relatively stable over time.

ecological carrying capacity. The ecologically determined maximum sustainable stock (e.g., animals, humans, etc.) that can be supported by a given unit of habitat.

ecological economics. A branch of economics that explicitly recognizes environmental limits, and the interactions between human activities and the environment. Ecological economics places a priority on the understanding of ecological processes, system, causes and effects in addressing economic and environmental problems.

ecological fallacy. The notion that analyses based on different levels of aggregation of data may give very different results.

ecological footprint. A measure of the appropriated ecosystem area that theoretically would be required to sustain a given population. Ecological footprinting is sometimes used as an indicator of the degree to which a population's consumption is ecologically sustainable.

ecological growth efficiency. The proportion of biomass energy produced at a trophic level that is used by the next highest trophic level.

ecological integrity. A term used to describe the general well-being of ecosystems in terms of habitat, species diversity and other factors.

ecological sensitivity. The potential carying capacity of an ecosystem for a given pollutant. For example, lakes found in areas where the soil is alkaline are less affected by acid rain than those found in areas where the substrate is acidic.

ecology. Study of life and interactions between organisms, and between organisms and their biotic and abiotic environment.

econometrics. A branch of economics that uses statistics to test hypotheses about economic models and to estimate model parameters.

economic bad. A good that is a useless byproduct of the production process. Many waste products are economic bads.

economic convergence. The notion that, over time, the per capita *levels* of income of different countries will converge to a common value. Conditional convergence is the notion that, over time, the *growth rates* in per capita income converge to a common rate.

FURTHER READING
Ray (1998).

economic cost. The full cost of producing a good or service including all explicit and implicit costs of the producer, and the costs or benefits absorbed by the rest of society.

economic deficit. The amount by which current spending exceeds current revenues.

economic development. Process by which individuals, communities and societies are able to become empowered so as to meet their needs in an equitable and efficient manner.

economic efficiency. Production with an optimal level and mix of inputs and outputs and which requires allocative, scale and technical efficiency. See *allocative efficiency, scale efficiency* and *technical efficiency*.

economic growth. A process that is usually defined as the sustained increase in the per capita real value of the production of goods and services in an economy.

economic instruments. Approaches to pollution control that involve economic incentives and include effluent charges, deposit-refund systems for recyclables, and market-based instruments. See *market-based instruments*.

economic rent. The return over and above that necessary to ensure the supply of a factor of production.

economic reserves. Those resources that have been identified (which includes actual deposits that have been measured as well as those that have been inferred) and can be extracted without prohibitive costs.

economics. Systematic study of aggregate human behavior and the decisions of individuals regarding their livelihood and well-being.

economic surplus.   The amount by which current revenues exceed current spending.

economies of scale.   The reduction in long-run average cost of production due to an increase in total output. Thus, if a firm triples its production but only doubles its costs, it benefits from economies of scale.

economies of scope.   A reduction in the total average costs of a firm by producing a range of related products rather than producing just one type of product.

economy.   Term given for the collective economic activities and interrelationships of a defined geographical region or country.

ecosphere.   The portion of the earth that includes the biosphere, hydrosphere, atmosphere, and soils and minerals.

FURTHER READING
Holmberg, Robert and Erikson (1996).

ecosystem.   A discrete or distinct unit or community of organisms with particular characteristics.

ecosystem approach.   An approach to environmental management that focuses upon the whole, and linkages among components within an ecosystem, including human activities, so as to ensure ecological integrity.

ecosystem services.   Term given to the services that the environment can provide as part of its natural functions. For example, the water cycle provides an invaluable service by helping to restore water contaminated with pollutants.

ecotechnological productivity.   The use of biotechnology to enhance the primary productivity of naturally occurring ecosystems for appropriation by the human economy.

ecotone.   A clearly defined zone that demarcates the boundary between two or more different communities.

ecotourism.   Tourism, travel, and adventure focusing on outdoor activities or outings that have as their primary destination the enjoyment of nature.   Often ecotourism is associated with attempts to promote tourism that is environmentally focused and environmentally sensitive. See also *wilderness tourism, adventure tourism* and *sustainable tourism*.

ecotoxicity.   The negative effects of a substance on ecological (and in some cases geophysical) systems.

ecotoxicology.   Study of the effect of chemicals on the functioning of an ecosystem, including all of its individual components.

ecotype.   A locally adapted variant of a population.

ecozones.   Method of classifying broadly similar terrestrial and marine ecosystems.

ecumene.   The area of the planet where there exists permanent human habitation.

edaphic.   The classification system in which vegetation type is determined primarily by soil and groundwater conditions.

efficient equilibrium.   An equilibrium where the marginal social cost of production equals the marginal social benefit. See *marginal social cost* and *marginal social benefit*.

efficiency.   See *Pareto efficiency, economic efficiency* and *statistical efficiency*.

efficiency wage.   The notion that firms may wish to pay more than the existing wage so as to elicit greater productivity from their workers.

effort-yield curve.   The production or output of renewable resources, such as the amount of fish caught, based upon the effort expended in trying to harvest the resource. In some cases, total yield first rises as effort increases and then falls despite increasing effort, such as when increased fishing efforts reduces stocks making it more difficult to harvest fish.

eigenvalues.   Also called the characteristic roots. See *characteristic equation* and *polynomial equation*.

El Niño.   An irregular warming of the surface waters of the eastern tropical Pacific which suppresses the cold ocean upwelling of the Humboldt current off Chile and Peru. An El Niño event can last up to two years and results in significant global changes in weather patterns including greater rainfall on the west coast of South America, droughts in Australia and New Zealand and drier and hotter summers in North America. The most recent El Niño event peaked in December 1997. See *La Niña*.

FURTHER READING
Allaby (1996).

El Niño-Southern Oscillation Event (ENSO).   See *El Niño*.

elastic.   Term used to describe the consumer or producer response to a change in price. A good is elastic if the percentage change in demand (or supply) equals or exceeds the percentage change in the price. For example, if the price of a good fell by one percent and demand increased by more than one percent, the good would have an elastic demand at that price. See *inelastic*.

elasticity.   A numerical measure of how responsive one variable is to changes in another variable. It is calculated as the percentage change in one variable divided by the percentage change in the other variable.

elfin thicket.   A densely vegetated grove composed of gnarled and stunted woody vegetation, usually found at high altitudes.

embedding.   Term used to describe the apparent insensitivity in contingent valuation surveys of respondents' willingness to pay measures for changes in environmental assets that may be "embedded" in each other. For example, the willingness to pay to ensure the continued existence of wolves in a national park may be very similar to the willingness to pay measure to ensure the continued existence of the park itself.

embodied technological change.   Technological change that arises from the use of new machines or capital goods. See *technological change* and *disembodied technical change*.

eminent domain.   The notion that the state has the right to appropriate private property for the common good provided that the owners are paid a fair market value for their property.

emission/effluent charge.   A pollution charge or tax levied on an emitter based on the amount of pollution emitted.

emission permit. A permit or property right which allows the holder to emit or release a given quantity of a pollutant in a defined period of time. Tradable emission permits allow different parties to sell and buy permits such that those who can meet the emission limits at low cost can sell some of their permits to others who are only able to meet their obligations at a high cost.

emission permit rental charge. A charge payable to a regulator that is based on the market price of emission permits.

FURTHER READING
Grafton and Devlin (1996).

emissions banking. The saving of emissions or pollution permits for future use. See *pollution permits*.

FURTHER READING
Tietenberg (1996).

emissions concentration. The level or amount of pollutants discharged into the environment measured in terms of volume.

emissions reduction credit (ERC). Term that refers to a verifiable reduction in greenhouse gas (GHG) emissions in one country, but which is financed by another. This process may allow the country financing the reduction to use the ERC when calculating its own level of GHG emissions. See *Kyoto Protocol* and *joint implementation.*

emission standard. A legal limit on emissions which, if exceeded, may result in a fine or other costs for the offending party.

emission trading. A system whereby firms or countries are allowed to purchase the right to emit or pollute from others so that they are able to meet their obligations, caps or limits on their own emissions. See *emission permit.*

endangered species. Plants and animals threatened with extinction including at least 3,000 animals and 24,000 plants listed under CITES. See *extinction* and *Convention on Intertaional Trade in Endangered Species.*

Endangered Species Act. Legislation passed in the USA in 1973 requiring the US government to identify species that are in danger of extinction. Species that are identifed are placed upon a list, at which point public and private landowners are prohibited from any activities that may harm the endangered species.

endemic disease.   Disease that is commonly found and continuously present in a particular region or location.

endemicity.   The degree to which a taxa or species is found only within the boundaries of a given area.

endemics.   See *endemic species*.

endemic species.   Species found in a defined geographical area.

endocrine.   Name given to the system, and its hormones, that is used to regulate body function in complex organisms.

endocrine disruptors.   Chemicals that affect the endocrine hormone system of animals, including organochlorines such as PCBs, DDT and many pesticides.   Such chemicals can lead to physical abnormalities and a variety of health problems.

FURTHER READING
Colborn, Dumanoski and Myers (1997).

endogenous growth.   Name given to economic models that try to explain economic growth endogenously such that growth is not explained by something exogenous such as technological progress. Such models have emphasized, among other things, the importance of human capital in economic growth.

FURTHER READING
Barro and Sala-i-Martin (1995).

endogenous variable.   A variable in a model or an equation which is determined by the model. See *exogenous variable*.

end-of-pipe treatment.   Methods of pollution control that focus on treatment of pollutants at the end of the production process rather than the prevention of pollution.

endothermic.   A process or reaction by which heat is absorbed from the surrounding environment.

energy budget. A measure of the energy flows within an economic or ecological system.

energy conservation. Methods and practices to reduce the levels of energy consumed.

energy crisis. Term used to describe the economic downturn and concern over the scarcity of non-renewable resources that followed the rapid increases in oil prices in 1973.

energy efficiency. The proportion of energy input into a system or process that can be transformed into useful work rather than dissipated.

energy intensity. Ratio of expenditures on energy to the total value of production in an economic process or system.

energy storage. Converting actual energy into potential energy that can be recovered later, e.g., using windmills to pump water into a reservoir when the wind is blowing so that the water can later be released to create power when there is no wind.

enforceability. The ability to protect and maintain the property rights of individuals, communities or firms.

Engel's law. So-called economic law observed by Ernst Engel that states that as income rises, the proportion of income spent on food declines.

ENGOs. Acronym for environmental non-governmental organizations.

entropy. The level of disorder in a system that, in a closed system, represents the amount of unavailable or bound energy. Living organims (including human beings) try to maintain, or at least slow down, an increase in their own entropy by using low entropy materials in their environment. Economic activity can be broadly characterized as the transformation of low entropy natural resources (such as oil) into higher entropy forms, such as wastes. See *second law of thermodynamics.*

FURTHER READING:
Georgescu-Roegen (1973).

enumerator. One who administers survey questionnaires.

envelope theorem. A mathematical theorem widely used in economics which states that the partial derivative of the optimal value function by a parameter is identical to the partial derivative of the associated Lagrangian function, holding all variables at their optimal values.

FURTHER READING
Grafton and Sargent (1997).

environment. The abiotic and biotic elements that form our surroundings.

environmental amenities. Utility enhancing attributes of the natural environment. Usually, amenities refer to non-extractive, consumption attributes (e.g., site attributes, air quality, views, etc.).

environmental assessment. In the USA, an assessment by the federal government of whether there will be any environmental impact from a proposed policy or action by a federal agency. If there is an impact, then an Environmental Impact Statement must be prepared.

environmental bond. A bond made by a firm (or an individual) against possible environmental damage that could result from that firm's actions. The firm can reclaim its bond by employing environmentally sensitive practices or through environmental remediation that leaves the environment at a level of quality agreed upon in the bond.

environmental covenant. A formal agreement between two or more parties in which the owner of an environmental asset or resource, in return for some consideration, agrees to restrict who has access and/or how the asset or resource is used. Covenants often arise between owners of land and governments interested in conserving the habitat or species found on private land.

environmental economics. The application of economic principles to the management of environmental quality. Environmental economics differs from resource economics in that it focuses on the valuation and management of environmental quality and not just resource stocks.

environmental equity. The degree to which environmental amenities and risks are distributed evenly within society.

environmental federalism. Term used to describe how in federal states (such as the USA, Canada, Mexico, Germany and Australia), policies have been developed to assign responsibilities to the appropriate level of government.

environmental impact analysis. See *EnvironmentalImpact Assessment.*

Environmental Impact Assessment (EIA). A detailed accounting of the impacts on the environment that would be expected given a proposed development. The findings of the EIA are presented in an environmental impact statement (EIS).

Environmental Impact Statement (EIS). The formal document that details the likely impacts on the environment due to proposed development. See *environmental assessment.*

environmental justice. The principle that poor people should not suffer the burden of environmental pollution and contamination disproportionately.

environmental Kuznets curve. An inverted U-shaped functional form relating levels of environmental degradation with income. The curve suggests that environmental degradation first increases as countries start to develop, reaches a turning point, and then subsequently declines as countries become wealthier.

FURTHER READING
de Bruyn (2000) and Ekins (2000).

Environmental Protection Agency (EPA). The US federal agency charged with controlling pollution.

environmental risk. Risks to human, ecological, or material well-being that emanate from environmental sources.

environmental services. The services provided by the environment which include the ability to assimilate and dispose of waste, provision of natural resources (such as fish and trees), and aesthetic and recreational value.

environmental services flows. A series of beneficial services produced by an environmental asset over time.

environmental sex lability. A characteristic of some plants and animals in which extremes in environmental factors (e.g., temperature or humidity) can determine sex. Environmental sex lability is known to occur in certain amphibians, reptiles and eels.

environmental shirking. When an individual or firm fails to comply with environmental regulations.

environmentally sustainable growth (path). A path of economic development in which per capita social welfare is monotonically increasing and environmental quality is non-decreasing over time.

environmental values. Human perception and satisfaction derived from the physical attributes or functions of a landscape or ecosystem.

enzyme. A set of proteins used by plants and animals to accelerate the speed of chemical reactions.

EPA. See *Environmental Protection Agency*.

Eocene. Geological epoch from 57.8 to 36.6 million years ago.

ephemeral. Short lived. Ephemeral refers to plants that thrive only for a short time, usually after heavy rain.

epidemic. A widespread infection or outbreak of a disease involving many individuals.

epidemiology. The study of human diseases using data on the distribution and size of diseases, and their causal factors.

epilimnion. A warmer and upper layer of water in a lake.

epiphyte. Plant that depends on other plants for physical support but not for nutrition.

epoch. A subdivision of a geological time period.

equatorial currents. Large surface currents found near the equator that flow in a westerly direction.

equilibrium. A state of balance in a system. In the static case, it represents a balance between competing forces (e.g., supply and demand). In the dynamic case, equilibrium represents the state of a system in which stocks tend to remain at constant levels over time.

equitability.   A measure of the evenness with which a resource or its associated benefit stream is distributed among members of society.

equity.   A term used to describe the belief that consideration should be given to the distribution of income and wealth within a society, in accordance with the identified needs of the members of socety.

equivalent variation.   A measure of consumer surplus based on the amount a consumer would be willing to accept to forgo a change in economic circumstances due to a proposed policy. The difference in consumer surplus is measured by substracting how much someone would have in the current situation, without the policy change, from what she would have under the proposed policy. This tells us what the individual would need to make her as well off without the policy change as she would have been with it. See *compensating surplus variation.*

ericoid.   Plants that demonstrate a shrub-like growth habit.

erratum.   Latin term for error.

error term.   See *residual.*

ESAM.   Expanded social accounting matrix. In addition to standard income and capital accounts, the ESAM also includes environmental stocks and flows.

escapement fishery.   A fishery that is managed by ensuring a certain number of fish are allowed to live or "escape" free from exploitation. Salmon fisheries are escapement fisheries as a certain number of salmon are needed to swim upstream to spawn and ensure the future of the fishery.

Escherichia coli.   Bacteria that live in the digestive tracts of vertebrates, including humans, and that provide a useful function in the absorption of nutrients. See *Escherichia coli 0157:H7.*

Escherichia coli O157:H7.   A recently discovered pathogen that can be found in the feces and tissues of some animals and can contaminate water supplies. The pathogen has been called "hamburger disease" because it was first observed in the United States and, in the initial outbreak, was transmitted from the sale of undercooked hamburgers. Outbreaks have occurred in several countries including Japan, England and Canada.

estuary.   The lower part of a river or stream that enters the ocean and is influenced by tidal waters.

ethane.   See *ethylene.*

ethanol.   Common name for ethyl alcohol. It can be produced from the fermentation of plants and can be used as a substitute for petroleum.

ethnobotany.   The study of the relationship between plants and people.

ethno-medicinal.   Traditional knowledge related to the collection and processing of naturally occurring medicines.

ethylene.   An important plant hormone that is emitted by ripe fruit and that is used commercially to hasten the ripening of stored fruit.

et sequens (et. seq.).   Latin term meaning "and the following".

eugenics.   The notion that the human species can be "improved" genetically by selective mating.

eukaryote.   Multicellular (including humans) and unicellular life forms whose cells are characterized by a nucleus and membrane and other features.

Euler's equation.   An equation that, under certain conditions, yields the necessary condition to maximize the value of the objective function in a dynamic optimization problem.

FURTHER READING
Chiang (1992).

euphotic zone.   Upper levels of a lake or the ocean where light is sufficient for photosynthesis.

European Union (EU).   A group of 15 European countries which share a common free trade area and a set of rules governing a range of industrial, labor and other policies. The EU grew out of the European Economic Community (EEC) established by the Treaty of Rome in 1957, which consisted of West Germany, France, Italy, the Netherlands, Luxembourg and Belgium.

eutrophic.   A body of fresh water which is highly productive because of high nutrient concentrations.

eutrophication. A natural process that can be hastened by human activity if it increases the level of nutrients and reduces the level of oxygen in the water. The runoff from farms which have received applications of phosphorous and nitrogen, in particular, can hasten eutrophication of small lakes and ponds.

evaporation. Process by which liquid in bodies (such as water in soil and oceans) is transformed into vapor or a gaseous state (such as water vapor).

evapotranspiration. An important part of the water cycle whereby evaporation of moisture at the earth's surface and transpiration from plants discharges water into the atmosphere. See *evaporation* and *transpiration.*

even-aged forest. Forest where the trees are of a similar age. Many commercial and artificial forests have even-aged stands of trees and it is claimed that this reduces ecosystem and species diversity relative to uneven-age and old-growth forests.

even-aged stand. A forest in which all stands are approximately the same age (within 20 years.) See *uneven-aged stand.*

evergreen trees. Tree species, such as cedar, that retain their leaves all year round.

evolution. The notion that changes in species occur incrementally over succeeding generations. See *Darwinian evolution.*

evolutionary facilitation. The notion that, over time, some species are becoming better able to evolve or adapt.

FURTHER READING
Wills (1989).

evolutionary paradigm. A mechanism for economic adaptation and learning using three basic processes: (1) information storage and transmission, (2) generation of new alternatives, and (3) the selection of superior alternatives according to performance criteria.

excess capacity. Capacity exceeding that required to produce a given level of output. See *capacity.*

excess demand.   Demand that exceeds the current level of supply at the existing market price.

excess supply.   Supply that exceeds the current level of demand at the existing market price.

exchange controls.   Controls imposed on individuals and companies to limit the exchange of one currency for another.

exchange rate.   The price of a currency in terms of other countries' currencies, or the rate at which one can exchange one currency for another.

excludability.   The ability to deny others access to, or use of, a good or asset.

exclusive economic zone (EEZ).   A zone of extended jurisdiction up to 200 nautical miles from the shores of coastal states.

exclusivity.   A characteristic that refers to how able a society, or the holder of a property right, is to exclude others from infringing upon recognized property rights. Without some degree of exclusivity, property rights do not exist.

exhaustible resource.   See *non-renewable resource*.

existence value.   The value placed on a good or resource solely because that good or resource exists (exclusive of any use value). For example, many people wish to ensure the continued existence of blue whales despite the fact they will almost certainly never see one or benefit from its "use". See *consumptive value* and *non-use value*.

ex officio.   Latin term for office or position (of an individual).

exogenous variable.   A variable in a model or an equation that is pre-determined that is, its value is determined outside of the model.

exothermic.   A process or chemical reaction in which heat is emitted into the environment.

exotics.   See *exotic species*.

exotic species.   Species that are not native to an environment. The introduction of exotic species, especially in isolated island habitats, has been a major cause of species extinction. See *native species*.

expansion path.   The path traced in terms of relative input usage from an expansion of output of a firm.

expected value.   The expected value of variable X is denoted by E(X) and is a weighted average of the variable, with the weights corresponding to the probability that the variable will take on a particular value. Thus if the probability that the variable X takes on a given value x_i is
$$Pr(X=x_i)=f(x_i)$$
and if X takes on n possible discrete values then
$$E(X)=x_1f(x_1) + x_2f(x_2) +...+x_n f(x_n).$$

expected value maximization.   A choice rule for consumption or resource allocation models in which the consumer or planner chooses among several uncertain (random) allocations of goods or resources in order to maximize the expectation of utilities from each possible outcome.   The expectation of utilities is calculated as the utility from each possible allocation times the probability of occurrence of that allocation.

expenditure function.   Function that represents the minimum level of expenditure necessary to achieve a defined level of utility.

exploitation rate.   The ratio of the weight of the fish harvested in a period of time to the total biomass of fish at that time.

explicit costs.   Costs that directly affect the cash flow of firms or individuals. See *implicit costs*.

exponent.   The power to which a variable is raised. For example, in the equation $b = a^3$, the variable a has an exponent of 3.

exponential function.   A function of the form $f(x) = b^x$ where b is the base and x is the exponent.

exponential growth.   Growth or increase which is a fixed percentage per time period. For example, a population which increases by 1.5 percent per year exhibits exponential growth.

exposure analysis.   A method of analysis that begins at the source of contamination and assesses its transmission and potential impacts on the environment, especially human health.

extant.   Species that currently exist.

extended jurisdiction.   The extension of national sovereignty outside of national borders, such as the creation of 200-nautical-mile exclusive economic zones. See *exclusive economic zone.*

extensive form.   See *game tree*

extensive margin.   The term for units of land where the value of production exactly equals the   costs of production. At the extensive margin, no differential rent is earned. See *Ricardian rent.*

FURTHER READING
Van Kooten and Bulte (2000).

external benefit.   The benefit that accrues to others but which is not accounted for in the actions of consumers and producers.

external cost.   The cost imposed on others and the environment that is not accounted for in the actions of consumers and producers.

external debt.   The debt owed by a country (both private and public debt) to foreigners.

external economies.   Positive externalities that arise in a region following an increase in production or output, such as reduced transportation costs due to a greater density of population.

external diseconomies.   Negative externalities that arise in a region or area following increases in production or output, such as traffic congestion.

externality.   A situation where the actions of an individual, company or community has an unintended effect on others or the environment, and these effects are not taken into account in the decision making of the parties causing them. For example, if a firm pollutes a river without considering the costs it imposes on downstream users, this represents an externality.

FURTHER READING
Baumol and Oates (1988).

extinction. Shortened form of species extinction, which has occurred throughout life on earth including several mass extinctions such as that which occurred at the end of the Permian period some 245 million years ago. The current rate of species extinction is considered by many observers to be several times higher than in the recent past. The main factors contributing to the high rate of species extinction are habitat destruction and species invasion where exotic species are introduced, often by anthropogenic action, into new habitats, negatively affecting indigenous species.

FURTHER READING
Kaufman and Mallory (1993).

extirpation. The eradication of a species in a particular area or region.

extractive use value. The consumptive value of a good when removed from nature.

extrapolation. Terms used for predicting future values of variables from past values.

extrinsic value. The value placed by a consumer upon an object derived from either the benefits the consumer receives or from interaction with the object, rather than from its intrinsic value, (e.g., the pleasure received from viewing wildlife rather than the inherent value such wildlife possesses). See *intrinsic value*.

Exxon Valdez. An oil tanker that ruptured its hull in Prince William Sound, Alaska in 1989. The oil spill led to billions of dollars in clean-up costs and damages.

FURTHER READING
National Geographic (1999).

F

F$_{0.1}$.   A constant rate of exploitation of a fishery that is slightly less than that which maximizes the yield per recruit. $F_{0.1}$ is a widely used but ad hoc method of determining the total allowable catch in a fishery.

factorial design.   A method of design in experiments that allows for statistical comparisons between different "treatments" or applications. For example, a factorial design may be used to assess the performance of different varieties of wheat in field trials where each wheat variety represents a treatment.

factors of production.   The inputs in the production process which, collectively, include land (natural resources and the environment), labor (physical, human and social capital), and capital (physical capital).

Fahrenheit scale.   Named after the seventeenth-century scientist G.D. Fahrenheit and still used in the United States. Under the Fahrenheit scale, 32 degrees is the temperature at which ice melts and 212 degrees is the boiling point of water. To convert a temperature from the Fahrenheit scale to the internationally used Celsius scale subtract 32 from the Fahrenheit temperature and then multiply by 5/9. See *Celsius*.

fair compensation.   The notion that appropriation of assets or resources by the state or another authority requires that those affected receive a compensation that reflects the value of the property. See *eminent domain*.

fair game.   A gamble or game where the cost of playing the game equals the expected return. See *expected value.*

fair market value.   The price of an asset or resource should it be sold in the current market. The term is often applied in real estate when assessing the value of property.

fallacy of composition.   The mistaken assumption that what might be true on a small scale must be true on a much larger scale. For example, an individual fisher may feel that increasing her catch slightly would have no effect on the fish population. However, if all fishers make the same decision, the aggregate effect might be enough to reduce the stock level, if not eliminate it.

falldown effect. The decline in timber harvest associated with the transition from a first cut on a mature natural stand of timber to a sustainable harvest-level with a balanced age–class structure.

fallow. Land that is left uncultivated that may or may not have been previously farmed.

false-negative rate. The proportion of results in a diagnostic test that fails to identify what is being tested for but is, nevertheless, present.

false-positive rate. The proportion of results in a diagnostic test that incorrectly identifies what is being tested.

falsifiable. Principle used in model-building that states that all the hypotheses derived from a model should be testable and can be proven false.

family. A classification of organisms by genera. For example, the family Hominidae includes the genus Homo, of which Homo sapiens sapiens (modern humans) is the only surviving species. See *genus*.

FAO. See *Food and Agricultural Organization of the United Nations*.

fauna. The animal life found in an ecosystem. See *flora*.

Faustmann rotation. The optimal time to cut a stand of uniformly aged trees assuming the land on which the trees are growing has no other value than the return it generates from timber and that the landowner plans to have an infinite number of rotations. The Faustmann rotation is shorter than the single-period rotation or Fisher rotation. See *Fisher rotation* and *Hartman rotation*.

FURTHER READING
van Kooten (1993).

FDA. See *Food and Drug Administration*.

fecal coliform (FC). See *coliform bacteria*.

fecundity. A measure of fertility commonly defined by the number of offspring per period of time.

feedback. An occurrence that arises when one component in a system changes in response to another component. A negative feedback arises when the response to a change in a component in a system tends to move the system back to its former state while a positive feedback tends to move the system away from the former state. For example, a positive feedback would be an increase in snow accumulation that increases the albedo of the earth and thus leads to more solar radiation being reflected from the earth that, in turn, tends to make temperatures even colder.

feedback loops. Processes in which the secondary effects tend to either reinforce the process (a positive feedback loop) or tend to limit the process (a negative feedback loop). An increase in the average birth rate in an animal population might tend to lead to a rapidly rising population at first as more adults lead to more breeding (a positive loop) until increased competition for food may lead to an increased mortality (a negative loop).

feedback mechanisms. Cyclical processes in system and climate change modeling in which process flows are indirectly self-influencing through an intermediary process. Feedback is considered positive (negative) when it increases (decreases) the level of flow in question.

feedback rules. Explicit rules that relate current decisions to the current state of nature. See *feedback.*

FURTHER READING
Grafton, Sandal and Steinshamn (2000).

feedlot. Pens or structures in which livestock are enclosed and fed in order to increase their rate of growth. Large feedlots can impose significant environmental costs if the animal wastes are not disposed of appropriately.

fee-simple. Land-ownership that is freehold and without encumbrances on transfer or inheritance.

fen. English term for a wetland, swamp or bog.

feral. Domesticated animals that have subsequently returned to living in the wild, such as dingoes in Australia.

fertility rate. The number of live births in a year per thousand women of child-bearing age in the population.

fertilizers.   Substances added to soil to aid in plant growth. Fertilizers include animal manure and compost as well as artificial substances like phosphates and nitrates. Too heavy a treatment of fertilizers can impose environmental consequences as the material may be washed or leached away and can contaminate water sources.

finfish.   Vertebrate fish species.

fire suppression.   A common practice in forest and park management in which forest fires are suppressed. A potential problem of fire suppression in forestry is that it increases the chances of very large fires by allowing combustible matter to accumulate. In parks it can change the nature of habitats and may be detrimental to some species.

first fundamental theorem of welfare economics.   An   economic theo-rem that states in a perfectly competitive market, with no externalities or public goods, the market equilibrium is Pareto efficient. The result provides much of the justification for the arguments in favor of "free markets". See *Pareto efficiency* and *second fundamental theorem of welfare economics.*

first law of thermodynamics.   See *law of conservation of energy.*

first-order conditions.   A set of necessary conditions that requires the first-order partial derivatives of the choice variables to be equal to zero if the objective function is maximized or minimized.

FURTHER READING
Grafton and Sargent (1997).

fiscal policy.   Any policy of the government that is related to its revenues and expenditures.

Fisher rotation.   Single-period rotation in forestry which states that the optimal time to cut a stand of uniformly aged trees is when the rate of growth in the value of the trees equals the rate of interest. See *Faustmann rotation.*

FURTHER READING
van Kooten (1993).

fishers' individual salmon harvesting rights (FISHRs).   Property rights for salmon fishers that allow them to bid for the privilege of participating at salmon fishing openings.

FURTHER READING
Grafton and Nelson (1997).

fishery management plans (FMPs). Plans developed by Regional Fishery Management Councils in the USA that set the broad objectives and strategies for managing specific fisheries.

FURTHER READING
National Research Council (1999).

fishing effort. A single measure that tries to aggregate the total effort or inputs used in fishing. Fishing effort may be defined as days at sea, fuel consumption, the number of fishing boats actively used, or any measure that is reflective of fishing inputs. Standardized fishing effort is a measure of fishing effort after making adjustments for differences in the gear and vessels used.

fishing mortality. Mortality, or death, of fish that is attributable to harvesting or capture. Fishing mortality and natural mortality together represent total mortality in a fishery.

fishing power. A measure of the ability of a fishing vessel to catch fish; may be a function of vessel characteristics, gear used, the skipper or the crew.

fission. See *nuclear fission*.

fitness. A measure of how well adapted an individual or species is to its environment.

fixed costs. Costs of production that do not vary with the level of output.

fixed effects. Term applied to statistical models where differences across units of observations can be identified by dummy variables or different constant terms. See *random effects*.

FURTHER READING
Greene (1997).

fixed exchange rate. A term that refers to when a country's exchange rate is fixed or pegged to that of other currencies. See *floating exchange rate*.

fixed factor. A factor of production that can only be changed in the long run, such as the size of a factory.

fixed input. See *fixed factor*.

flexibility. A property right characteristic that refers to the ability of the holder of the property right to accommodate changes in the asset and individual circumstances.

flexible functional forms. Functions used to estimate costs, profits and output from data, that do not impose arbitrary restrictions on the elasticities of substitution among inputs or outputs. A commonly used flexible functional form is the Translog.

FURTHER READING
Takayama (1985).

floating exchange rate. A term that refers to when a country's exchange rate is able to adjust freely to changes in currency markets. See *fixed exchange rate*.

flocculation. Term used to describe the aggregation of fine sediments suspended in a solution. Flocculation is used in sewage treatment to help remove fine particles of waste from treated water.

flood plain. The area of land surrounding a water channel that is subject to flooding.

flora. The plant life found in an ecosystem. See *fauna*.

flow. The input or output from a stock that can vary over time. The annual harvest of fish from a fishery is a flow. See *stock*.

flow pollutant. A pollutant that imposes social costs at the time of its discharge, but does not accumulate in the environment, such as noise pollution.

flower. The part of an angiosperm used for reproduction. See *angiosperm*.

flue gas desulfurization. Method for removing some of the sulfur dioxide produced in the combustion of coal by the spraying of the fumes in a discharge chamber with a lime-rich liquid so as to produce calcium sulphate.

fluidized bed combustion. Method for reducing sulfur dioxide emissions by burning coal with a mix of limestone and sand.

fluoridation. See *fluoride*.

fluoride. A form of the element fluorine that, in small doses, can be beneficial to the development of teeth and bones. Many urban communities add small quantities of sodium fluoride to drinking water for its potential health benefits, but this procedure is becoming increasingly controversial as some individuals are unable to tolerate even small doses (one part per million).

F_{max}. A constant rate of exploitation of a fishery that maximizes the yield per recruit.

folivorous. Leaf-eating.

folk theorem. In game theory, the hypothesis that as the time horizon approaches infinity, the incentive to "free-ride" will diminish.

Food and Agriculture Organization (FAO). An organization of the United Nations based in Rome, Italy, that is mandated to improve the living standards and nutrition of rural populations.

Food and Drug Administration (FDA). US federal agency charged with ensuring the safety of food and drugs consumed within the United States.

food chain. The chain of events in the food consumption of an ecosystem in which plants are eaten by animals which in turn are eaten by larger animals.

food web. See *food chain*.

foot. Unit of length equal to 0.3049 of a meter.

foehn. Dry wind that blows on the eastern side of a range of mountains.

forage. Vegetation, including grasses, herbs, shrubs, and leaves that can be used to feed livestock.

Fordism. Term describing the large-scale mass production of goods, often characterized by assembly lines.

forest management. The organized management of forest stands to achieve stated outcomes. Forest management often refers to a plan of use and stewardship, as implemented by a private owner or public agency proprietor.

forest principles. Principles of forest management agreed to at the 1992 United Nations Conference on Environment and Development.

forest profile. A description of the range of forest conditions that exist across a given landscape. A forest profile may contain information on timber species, condition, age, location, elevation, and so on.

forestry. The science and practice of managing forest resources for human benefit.

formaldehyde. A poisonous gas that is widely used in plastics and that can be a major source of indoor air pollution as it is slowly emitted from such products as laminated woods, upholstery and even permanent-press fabrics.

forward market. Any market in which buyers and sellers contract to buy or sell goods in the future at fixed prices.

fossil. The geological remains or evidence of past life.

fossil fuels. Sources of energy, such as oil, natural gas and coal that are obtained from organic matter that was laid down millions of years ago.

fractals. Seemingly complicated patterns where the degree of irregularity is the same at any scale. Fractals can be formed from equations that exhibit chaos. See *chaos*.

fragility. The opposite of resilience. See *resilience*.

Framework Convention on Climate Change (FCCC). A convention signed by over 160 countries at the 1992 Earth Summit in Rio de Janeiro and which came into force in March 1994 following its ratification by over 50 nations. Annex I countries to the convention committed themselves to return their emissions to 1990 levels by the year 2000. This commitment was modified in the 1997 Kyoto Protocol. The ultimate objective of the FCCC is to stabilize greenhouse gas concentrations at a level that would prevent dangerous anthropogenic interference with the climate. See *Kyoto Protocol.*

free access.   See *open access.*

free good.   Goods that are so abundant they have no market price. Paradoxically, free goods can be the most essential, such as the air we breathe.

free radical.   Name given to any atom or molecule that has a "free" electron and is thus highly reactive.

free-rider.   An individual who benefits from the actions and efforts of others without contributing to the costs incurred in generating the benefits.

free trade.   The unimpeded international exchange of goods and services without trade restrictions such as tariffs, quotas and non-tariff barriers.

freon.   Common name for a chlorofluorocarbon. See *chlorofluoro-carbons.*

frequency distribution.   An approach to representing data where a variable is classified into appropriate groupings, levels or classes with the frequency of the number of observations in each grouping.

fresh water.   Water (H_2O) that contains salt in quantities of less than 0.05 percent.

FURTHER READING
Pielou (1998).

freshwater ecology.   Limnology. The study of aquatic systems in fresh waters such as lakes, ponds, rivers, streams, and other water bodies.

front-end charge.   Charge paid by consumers upon purchasing a product. The charge may be refundable if the waste product is disposed of appropriately.

frugivorous.   Fruit-eating.

F-test.   A frequently used statistical test for the equality of variances. In regression analysis, an F-test is a joint test used to test the null hypothesis that all coefficients are equal to zero with K–1 and n–1 degrees of freedom, where K is the number of regressors and n is the number of observations.

FURTHER READING
Greene (1997).

fuel loads.   Accumulated dead wood, fallen timber and other combustible material found within a forest stand. More of these materials means a higher fuel load, and an increased likelihood that fires will be severe.

fuel switching.   The switching or substitution of one fuel for another. For example, fuel switching from the use of coal to natural gas in the generation of electricity reduces emissions of carbon dioxide for a given amount of electricity produced.

fuelwood.   Wood that is harvested and consumed for heating and cooking purposes rather than being converted into other products.

fugitive resource.   A natural resource that is non-stationary, such as a fishery.

full-cost pricing.   The pricing of goods and services such that the price the final consumer pays includes both the private and social costs of its production. See *social cost* and *externality*.

full-tree harvesting.   A method of timber harvesting where the trunk, branches, and sometimes roots are removed at the time of harvest. Full-tree harvesting is sometimes employed to control root disease.

fully regulated forest management.   Management of a forest where the forest is divided into a number of areas each capable of producing an equal volume of timber and where the total number is equal to the length of time in years in which it takes the trees to reach the age at which they can be harvested. Each year, one area or stand is harvested.

function.   1. A relationship whereby the independent variables determine unique values for the dependent variable.   2. Activity of an organism.

fundamental equation of renewable resources.   The equation that defines an optimal steady-state for the level or size of a renewable resource. In its general form, the fundamental equation may be defined as, $\rho = F'(x) + \dfrac{\partial \pi / \partial x}{\partial \pi / \partial h}$

where ρ is the discount rate, x is the level or size of the renewable resource, h is the harvest, π is the current rent from the renewable resource, $F(x)$ is the growth function of the renewable resource, $F'(x)$ is the change in growth from a change in the size of the renewable resource and the second term on the right-hand side of the equation is called the marginal stock effect. The left-hand side of the equation may be interpreted as the external rate of return (return on assets separate from the renewable resource) while the right-hand side may be interpreted as the internal rate of return.

FURTHER READING
Conrad (1995).

fundamental equation of non-renewable resources. See *Hotelling's rule*.

fund pollutant. See *stock pollutant*.

fungi. Micro-organisms that develop spores to reproduce and include mushrooms, yeast and molds. Unlike plants or algae, fungi cannot photosynthesize and must absorb their nutrients from their surrounding environment.

furans. Chlorine-based chemical compounds that have similar health effects to dioxins. See *dioxins*.

fusion. See *nuclear fusion*.

futures. Contracts to deliver or buy goods in a futures market at a defined price.

future value. Value of returns assessed for a particular period in the future.

fuzzy logic. Methodology used to model problems that involve fuzzy sets. See *fuzzy sets*.

fuzzy sets. Sets whose membership or elements are uncertain. Fuzzy sets are used to model situations where the degree, or the extent to which an event occurs is uncertain.

G

G7/G8. G7 is a grouping of the world's largest industrial economies consisting of the USA, Japan, Germany, France, Italy, the United Kingdom, and Canada. The G8 includes the G7 countries plus Russia.

G77 countries. Grouping of developing countries.

Gaia hypothesis. Idea popularized by James Lovelock, an atmospheric scientist, who hypothesized that the earth could be viewed as a single organism where its constituent elements coexist in symbiosis. Thus animals and plants collectively help to manage a world, through positive and negative feedbacks, and ensure the planet maintains an equilibrium that is desirable for all life. See *daisyworld*.

FURTHER READING
Lovelock (2000).

gallery forest. Tropical forests that grow at high densities and tend to follow riverine courses. See also *riparian*.

gambler's fallacy. A belief that because an event has not occurred for a long period of time then it is more likely to occur in the near future.

game ranching. Raising wildlife under controlled conditions within pens.

game theory. An area of study and research that deals with the strategic interactions of (utility) maximizing agents under specified sets of constraints.

game tree. Also known as extensive form of a game, this is a structure that provides information about a game where the nodes represent potential moves in the game and the edges or lines or branches that move from the nodes represent the possible strategies or actions available at a particular node. Each node is labeled for the player making the move or, if the node has no edges or branches, it is labeled with the pay-offs for the chosen actions. See *game theory*.

FURTHER READING
Binmore (1992).

gamma radiation. The most dangerous form of the three main types of nuclear radiation. It consists of a stream of particles released from radioactive isotopes. See *alpha radiation* and *beta radiation*.

gap analysis (GA). Analysis designed to bridge or fill a gap in either knowledge or a defined objective. For example, in the USA, gap analysis includes various techniques to identify areas of importance in terms of biodiversity and then to give priority for conservation of areas that are not currently under protection.

garbage in, garbage out. The notion that if the data used in an analysis is of poor quality then the results will also be of poor quality.

gas flaring. The practice of burning off natural gas found at oil well sites.

gasohol. A mix of gasoline, ethanol and methanol.

Gaussian distribution. See *normal distribution*.

Gauss–Markov theorem. A theorem that proves that the least squares estimator of the parameters of a multiple regression is the best linear unbiased estimator (BLUE) with lowest variance, provided that the residuals are identically distributed with a mean of zero.

FURTHER READING
Ramanathan (1992).

GDP. See *gross domestic product*.

gear restrictions. Limits placed on the type of gear that may be employed in the harvesting or capture of fish and other animals. Gear restrictions are a form of input control and are widely used in fisheries management.

Geiger counter. Shortened name for the Geiger–Muller counter that measures ionizing radiation by assessing the electrical field disruption of the argon gas in the instrument.

GEMS. See *global environmental monitoring system*.

gene. A unit hereditary factor comprised of a segment of deoxyribonucleic or ribonucleic acid that codes for one or more related functions of an organism.

gene bank. A repository of plants, seeds and genetic material stored for the purpose of preserving genetic diversity.

gene pool. The total number of genes possessed by the reproductive members of a population.

General Agreement on Tariffs and Trade (GATT). An international trade agreement that existed from 1948 until 1994 when the GATT was replaced by the World Trade Organization (WTO). The GATT's objective was to limit or eliminate restrictive trade practices.

general circulation model (GCM). Models of the earth's climate which include several components. GCMs have been widely used to assess the potential impacts of increased concentrations of greenhouse gases (GHGs) on the future climate.

general equilibrium analysis. A method of economic analysis which attempts to simultaneously link all sectors and markets of an economy. See *partial equilibrium analysis*.

generalized least squares (GLS). A method of estimating parameters of a statistical model where there exists autocorrelation or heteroscedasticity and where ordinary least squares (OLS) would lead to inconsistent estimators. See *ordinary least squares*.

FURTHER READING
Greene (1997).

generation. 1. The predetermined "life span" of an agent in an economic or biological model. 2. A cohort of a population of a similar age.

genetically modified (GM). Term usually applied to plants where one or more of their genes have been altered by genetic engineering for a particular purpose. Genetically modified foods have caused a great deal of controversy due to concerns over the potential long-term health hazards.

genetically modified organisms (GMOs). See *genetically modified*.

genetic diversity. Variablity in genotypes found within a species or population. See *biodiversity*.

genetic drift. Chance fluctuations, independent of natural selection, in the genetic make-up of an isolated population.

genetic engineering. The anthropogenic change in genetic structure for a desired purpose brought about by the removal of, or a change in, chromosomes.

genetics. The science and study of how genetic influences affect organisms and the inheritance of characteristics.

genome. The genetic material of an organism contained in its chromosomes.

genomics. The study of genomes.

genotype. The specific genetic composition of an organism. See *phenotype.*

gentrification. The purchase and renovation of homes in run-down urban neighborhoods, often displacing lower-income households.

genuine progress index (GPI). An index of overall societal well-being that includes attempts to take into account changes in social, human, physical and natural capital to reflect the progress, or lack thereof, in a society.

genus. (Plural genera). A classification of species with common characteristics. For example, modern humans (Homo sapiens) are the only surviving species of the genus Homo. See *species.*

geochemically scarce metals. Those metallic elements the relative abundance of which falls while costs of extraction rise because of a mineralogical threshold.

geoengineering. Anthropogenic attempts to change global environmental systems. For example, some scientists have proposed placing reflective material high in the earth's atmosphere to increase the albedo and thus lower temperatures in response to possible global warming due to increased concentrations of greenhouse gases (GHGs) in the atmosphere.

geographic information system (GIS). Digitalized approaches to combining and analyzing information and data to visualize existing data.

geology.   The study of the earth, its structures and processes.

geometric growth.   Growth or increase where the ratio of the next value in the series to the current value is the same as the ratio of the current value to the previous value. For example, the series 3, 6, 12, 24 ... exhibits geometric growth where the ratio of the next number in the series to the current number is always 2.

geometric mean.   A summary statistic defined as the nth root of the product of a series of n numbers. For example, the geometric mean of the numbers 3 and 48 is $(3 \times 48)^{1/2} = 12$.

geometric series.   A series of numbers that can be written as
$$a + ac + ac^2 + \ldots + ac^n$$
given c does not equal 1. If $-1 < c < 1$ then the sum of an infinite geometric series is $a/(1-c)$.

geothermal.   Heat sources from beneath the earth's surface. In Iceland, geothermal energy is an important source of the total energy used.

geostationary.   Description of a satellite that does not appear to move its position relative to the planet around which it orbits.

ghost fishing.   The capture of fish by fishing gear that has been lost or abandoned.   For some gear, such as driftnets, ghost fishing can be a major cause of fishing mortality.

giardia.   A water-borne parasite that infects animals, including humans.

giga.   Term signifying a billion or 1×10^9.

gill net.   A fishing net that ensnares the heads of fish and that is attached to buoys at the surface.

Gini coefficient.   A frequently used measure of income and wealth inequality.   The Gini coefficient measures the area between the Lorenz curve and the 45° line relative to the entire area below the 45° line, where a measure of 0 indicates that all individuals have identical incomes. See *Lorenz curve.*

FURTHER READING
Ray (1998).

GIS.   See *geographic information system.*

glaciation. Cooling that is associated with the accumulation of snow and ice.

glacier. A land-bound body of ice formed by the accumulation of snow in excess of snow melt. In past ice ages, glaciers covered a significant part of North America and Europe.

global commons. Aspects of the environment that are global in nature and are common-pool resources, such as the earth's atmosphere. See *common-pool resource.*

global ecological services. Global benefits provided by aspects of the environment. For example, growing forests provide a global benefit by sequestering carbon.

global environmental facility (GEF). A fund established by the United Nations but held at the World Bank. The GEF is used for projects and initiatives in developing countries to prevent or mitigate global environmental problems.

global environmental monitoring system (GEMS). A system administered by the United Nations Environment Program. The GEMS monitors pollution, particularly airborne pollutants.

global positioning system (GPS). A system whereby a receiver can be used to track particular satellites in the earth's orbit to determine a spot on the earth's surface within meters of where the receiver is located.

global village. Term used to describe a set of ideas that represents the interconnectedness of people across the globe for good (communications, transport, etc.) and ill (spread of disease, global pollution, etc.).

global warming. The hypothesis that the anthropogenic emissions of greenhouse gases are leading to an enhanced greenhouse effect that will raise the earth's average surface temperature. The United Nation's International Panel on Climate Change estimates that the earth's average temperature will rise between 0.1–0.2° C/decade over the next 50 years (IPCCs 1997).

FURTHER READING
Houghton (1997).

global warming potential. An index which assigns values to greenhouse gases (GHGs) based on their estimated relative potential contribution to global warming. Carbon dioxide is assigned a value of 1 and gases with a value greater (less) than 1 (methane has a value of 21) are expected to have a greater (lower) relative contribution for an equivalent amount emitted.

GNE.   See *gross national expenditure.*

GNP.   See *gross national product.*

golden rule.   Term used to describe a time path for an economy where the level of savings and investment maximizes long-term per capita consumption.

Gondwanaland.   A supercontinent that broke off from the even larger supercontinent Pangea and eventually split and formed into Antartica, Africa, India, Australia, New Zealand, Sri Lanka and Madagascar.

goodness of fit.   A measure of how well the predicted values from a statistical model "fit" the observed values from sample data. A commonly used measure is the R-squared of a model.

Gordon–Schaefer model.   A bioeconomic model of a fishery that combines a surplus production model, or biomass dynamic model, for the population dynamics of the resource, and a static profit maximizing model of fisher behavior. A fundamental result of the model is that the maximum economic yield occurs at a higher level of the resource stock than the maximum sustainable yield.

governance.   The structure and practice of decision making in an organization or society.

government failures.   Errors of commission and omission for which the government is directly responsible. For example, a nation that is unable to provide security of property rights and personal security for its citizens suffers from government failure. See *market failure.*

GPI.   See *genuine progress index.*

Granger test for causality.   Test used to help determine the direction of causality between two variables. If a variable X "Granger causes" a variable Y, then past values of X are useful in predicting future values of Y.

FURTHER READING
Granger (1969).

grassland.   Biomes, such as prairies or savanna, that consist primarily of grasses. Most grasslands have been modified by human activity.

gravity model.   Model based on Newton's Law where the gravitational force between two masses is proportional to the product of their masses and inversely proportional to the square of the distance between the masses. Modified forms of the model have been used to help explain other phenomena such as trade between countries and migration between geographically separated populations.

Gray (Gy).   The internationally accepted measure for absorbed ionizing radiation where 100 rad = 1 Gy.

gray water.   Water which does not contain excreta and which is normally disposed of through the stormwater system.

grazing lease.   A lease on land for the purpose of grazing livestock.

grazing pressure.   The reduction in vegetative cover on a land parcel due to grazing by livestock.

Great Lakes Water Quality Agreement (GLWQA).   A 1972 agreement (up-dated in 1978) between the USA and Canada to reduce the pollution entering the Great Lakes (Superior, Michigan, Huron, Erie and Ontario) which are shared between the two countries. See *International Joint Commission.*

green accounting.   Approaches to national income accounting in which an attempt is made to record changes in the quality of the environment in physical and/or value measures.

FURTHER READING
Nordhaus and Kokkelenberg (1999).

greenbelt.   An area of natural habitat, largely free from development, set aside by easement or by community designation, as a buffer to future development or as a conservation area.

greenfield site.   Agricultural land that is highly desirable for conversion into a developed site for industry or residential purposes.

green GDP. The conventional measure of gross domestic product adjusted for changes in values of the environment.

greenhouse effect. Refers to the process whereby radiation from the sun passes through the atmosphere and warms the earth's surface. In turn, the earth's surface retransmits some of the energy from the sun as infrared or thermal radiation. Greenhouse gases (such as carbon dioxide) in the atmosphere are able to absorb some of the earth's thermal radiation thereby preventing the heat from leaving the planet. The greenhouse effect occurs naturally and without traces of greenhouse gases the earth's average surface temperature would be about –6 deg. C instead of its actual value of around 15 deg. C (Houghton, 1997). The enhanced greenhouse effect refers to the increased proportion of greenhouse gases in the atmosphere due to human activity. See *greenhouse gases*, *global warming* and *climate change*.

FURTHER READING
Houghton (1997) and Somerville (1996).

greenhouse gases (GHGs). Molecules in the atmosphere which are able to absorb the thermal radiation emitted from the earth. The most important greenhouse gases include carbon dioxide, water vapour, methane, nitrous oxides and chloroflurocarbons. See *greenhouse effect.*

green revolution. Period beginning in the late 1940s whereby plant scientists were able to use plant breeding to develop much higher-yielding varieties (HYVs) of various grains such as rice and wheat. The subsequent diffusion of the HYVs in countries such as India and Pakistan, along with the increased use of fertilizers and pesticides, significantly increased world grain production.

green-tree retention. A method of harvest in which live trees of specific species and size are reserved (spared) during harvesting.

grim strategy. The behavior or set of actions in a game chosen by a participant to impose the worst possible outcome upon the other participants, regardless of the effect upon themselves.

gross domestic product (GDP). A measure of the total value of goods and services produced in an economy in a year.

gross national expenditure (GNE). A measure of the annual total expenditure in an economy including consumption, investment and government expenditure.

gross national product (GNP). A measure of the total value of goods and services produced in an economy in a year, including net income paid to foreigners less net income paid to domestic nationals.

gross primary production. The gross energy produced by organisms before respiration.

gross registered tons (GRT). A measure of the size of a vessel that is listed on its registration certificate.

gross world product. The total value of goods and services produced by all economies over one year.

groundfish. Term usually applied to demersal fish species (such as cod) that are commonly found near the sea bottom.

ground-level ozone. See *tropospheric ozone.*

ground resolution element. The smallest unit of the earth's surface that can be detected from a remote sensing devise.

groundtruthing. Checks on the reliability of data obtained from secondary sources, or from remote sensing, by verifying the original sources of data or through direct observation.

groundwater. Water contained in rocks and soil beneath the earth's surface and which has precipitation as its source of replenishment. The rate of withdrawal of many groundwater supplies or aquifers exceeds the rate of recharge.

Groves–Ledyard mechanism. A scheme for eliciting a level of a public good that is Pareto efficient whereby individuals report to a planner the increment (which can be positive or negative) of the public good they would like, in addition to the total amount demanded by everyone else.

FURTHER READING
Cornes and Sandler (1996).

growing season. The span of time during the year in which plants are actively increasing in size.

growing stock. The volume of merchantable trees that are still actively growing within an area.

growth overfishing.   Term used to describe how a reduction in the total fishing effort can increase the total future harvests by increasing the average weight of fish.

growth rate.   A measure of change in a variable. Using discrete units of time for a variable x, the growth rate between period t-1 and t is
$$[(x_t - x_{t-1})/ x_t] \times 100.$$

Gulf Stream.   A warm ocean current named after its place of origin, the Gulf of Mexico, which passes up the coast of eastern North America past Iceland and eventually reaches the waters around Scandinavia. Without the Gulf Stream, Western and Northern Europe would be much colder.

quantification step. Because this is the first step, error at this point can lead to major problems and can also in some cases be reduced by increasing the relevant sample size.

purchase. A measured change in a variable. Using a series of alternating moves, the growth rate between prices i and $i+1$.

wild fluctuation. A wild swing in one or more variables of output, the cost of stocks. Such fluctuations in the cost of certain items. Another issue is how to quantify this. Perhaps the sample size fluctuations. Allowing for quantification and the degree that a change is small to be fully curbed.

H

habitat.   The environment in which a given plant or animal is usually found. For example, gazelles are usually found in a savanna habitat.

Habitat Conservation Plan (HCP).   In the USA, a plan prepared by landowners that specifies how they will deal with an endangered species found on their land. HCPs are submitted by landowners who hold permits for the "incidental taking" of endangered or threatened species on non-federal lands during otherwise lawful activities. The permits and HCP form an agreement between the landowner and the federal government that allows the landowner to engage in activities that might result in the landowner affecting the endangered species or modifying its habitat in exchange for measures that will enhance the survival of the species. See *incidental taking* and *Endangered Species Act.*

habitat diversity.   Variations in the ecology of a site that increase the number of niches for different species.

Hadean era.   A geological era that started at the formation of the earth 4,550 million years ago and continued until life first began some 3,800 million years ago.

Hadley cell.   A circulation system whereby rising warm air near the equator moves towards higher latitudes where the air cools and sinks back towards the subtropics, eventually returning to the equator.

half-life.   The time it takes for half of the material of a radioactive substance to decay.

halocarbons.   A family of compounds containing carbon and either bromine, chlorine or fluorine.   These can act as greenhouse gases (GHGs) in the atmosphere. Chlorofluorocarbons (CFCs) are halocarbons and can damage stratospheric ozone. See *chlorofluorocarbons.*

halogens.   Term for a group of highly reactive elements that includes chlorine, fluorine, bromine and iodine.

halons.   Organic compounds, formed from bromines and fluorine, which have been found to damage stratospheric ozone.

halophyte.   Plant adapted to growing in a salt-rich environment.

Hamiltonian.   An expression that includes the function to be optimized over time in a dynamic optimization problem plus the dynamic constraint multiplied by the co-state or auxiliary variable. Using the Hamiltonian, the necessary conditions for a maximum can be derived with optimal control theory.

FURTHER READING
Chiang (1992).

Hammer–Aitoff Conic Equal Area Projection.   A   type   of   map projection used in AVHRR and NDVI data. See *AVHRR* and *Normalized Difference Vegetation Index*.

hantovirus.   An often fatal viral disease spread by rodents and exposure to their excreta. The spread of the disease to different regions has been aided by the unwitting transport of rodents in cargo ships.

hard energy.   Energy provided from large-scale production facilities using non-renewable resources such as uranium, coal or oil. See *soft energy*.

hardpan(s).   A hard, usually impermeable, sub-surface soil layer often created by the precipitation of iron and aluminum.

hard water.   Water that contains a high proportion of alkaline compounds derived from calcareous sources. See *soft water*.

hardwoods.   Any member of the class of deciduous, broad-leaved trees (such as oak or maple), especially those that are suitable for timbering.

harmonic mean.   A summary statistic defined as the arithmetic mean of the reciprocal of a series of numbers. For example, the harmonic mean of the numbers 3 and 6 is $\frac{1}{2}(1/3 + 1/6) = \frac{1}{4}$. See *arithmetic mean*.

Hartman rotation.   The optimal time to cut a uniformly aged stand of trees assuming the trees provide both timber and non-timber values. The Hartman rotation is a modification of the Faustmann rotation and includes amenity and non-timber values. Provided that non-timber values increase with the age of the trees, the Hartman rotation is longer than the Faustmann rotation. See *Faustmann rotation*.

FURTHER READING
van Kooten (1993).

Hartwick's rule.   The rule requires that the rents from the extraction of non-renewable resources be invested in man-made capital, rather than consumed, so as to ensure a sustainable level of consumption over time.

FURTHER READING
Hartwick (1977).

harvesting.   The anthropogenic removal of biomass from an ecosystem. Harvesting of renewable resources without well-defined property rights or controls can lead to biological and economic overexploitation.

harvesting costs.   The fixed and variable costs associated with harvesting a natural resource. See *natural resources*.

hazardous pollutants.   Material discharged into the environment which can cause harm to humans within localized areas.

HCP.   See *Habitat Conservation Plan*.

healing.   The act of improving an individual's health (which could include physical, mental, or spiritual well-being).

health capital.   A measure of health or health status that can depreciate over time and can be maintained or improved over time through investment (e.g., exercise).

FURTHER READING
Grossman (1972).

health production function.   A numerical production function in which the health of an individual is the product of a combination of exogenous factors (e.g., environmental variables) and choice variables (health care, preventative medicine, etc.).

healthy-worker effect.   A widely observed phenomenon whereby employed workers often have lower mortality and morbidity rates than other groups in a population.

heavy metals.   Metals of a high atomic weight including lead, mercury and cadmium. These metals, and in their associated soluble organic forms, can pose significant environmental and human health hazards.

heavy water. Water in which normal hydrogen atoms have been replaced by hydrogen atoms with twice the atomic weight.

Heckscher–Ohlin model. Model of international trade in which for which the countries export goods whose factors of production are locally abundant.

hectare. Unit of area equal to 10,000 square meters or 2.47 acres.

hedonic analysis. Valuing activities or uses based on the observable characteristics or attributes of a good or service, such as valuing the effects of noise pollution by using property prices near and far away from an airport. See *hedonic property model*.

hedonic methods. Empirical methods used to estimate the implicit or hedonic prices of goods, or attributes of goods, using the explicit or market prices of goods. For example, the implicit price of air quality in a neighbourhood may be valued using data on property prices, house characteristics and environmental characteristics.

FURTHER READING
Rosen (1974), Griliches (1971) and Callan and Thomas (2000).

hedonic pricing. A method of evaluating the economic value of goods based on econometric techniques that identify and quantify the utility of various attributes of the good.

hedonic property model. A (hedonic) model of preferences for property characteristics in which the (resale) value of property is modeled as a separable function of the attributes of that property.

hedonic travel cost model. A travel cost method, first developed by Brown and Mendelsohn (1984), in which the implicit value of recreational site attributes are thought to contribute to the total revealed willingness to pay to gain access to a recreational site, as evidenced by an individual's choice to pay for the travel costs to a recreational site.

hedonic wage model. A hedonic model of preferences for job attributes, including job-related risks, in which the wage is modeled as a separable function of job and worker attributes.

height–diameter relationships. The correlation between the diameter and height of different tree species in terms of either growth and/or its eventual mature size.

Helsinki Agreement. A 1989 modification to the 1987 Montreal Protocol in which parties agreed to end the production of chlorofluorocarbons (CFCs) by the year 2000 so as to prevent further deterioration of stratospheric ozone.

Helsinki Process. A meeting between European countries called the Second Ministerial Conference on the Protection of Forests in Europe, held in Helsinki in 1993, and which identified the general guidelines for sustainable forestry management in Europe and the development of a series of criteria and indicators. See *Montreal Process*.

hepatitis. A viral disease classified by three types A, B and C. Hepatitis A is transmitted through contaminated water and food that is often a result of poor hygiene and inadequate sanitation.

herb. A seed-bearing plant that does not have, and does not develop, wood tissues.

herbicides. Chemicals used to inhibit growth or destroy undesirable plants. Herbicides are widely used, and their misuse can lead to serious environmental problems including groundwater contamination.

herbivores. Animals, such as cows and horses, whose diet consists primarily of plants.

heredity. Transfer of genetic traits to offspring. The degree to which offspring exhibit the physical traits of their parents depends on the inheritance of the characteristic, environmental factors and random factors.

Hessian matrix. A square matrix that consists of the second-order partial derivatives of a function, $f(x_1, x_2, \ldots, x_n)$, where the element in row i and column j is is the second-order partial derivative f_{ij}. The Hessian matrix can be used to determine the curvature of a function.

heteroscedasticity. A statistical term for error terms or residuals that are correlated with one or more of the variables in the model. Heterscedasticity commonly occurs with cross-sectional data and, because it affects the standard errors of estimators, it invalidates hypothesis tests about the estimators. If heteroscedasticity is present, generalized least squares (GLS) may be used.

heterotroph. An organism that obtains its nutrition from other organisms. See *autotrophy*.

hexachlorobutadiene (HCBD). A byproduct of the manufacture of some organochlorines which can bioaccumulate in the environment.

Hicksian demand. Sometimes called compensated demand. The Hicksian demand for a good or service is a function of the prices of goods and the level of utility to the consumer. By contrast, the normal or Marshallian demand is a function of prices and the level of income of the consumer.

Hicksian income. Named after the Nobel Laureate economist, Sir John Hicks, it is the national income that can be consumed while maintaining a non-decreasing level of the capital stock.

Hicks neutral technological change. Disembodied technological change where the ratio of capital to labor in production remains unchanged over time. See *technological change* and *disembodied technical change*.

high forest. Old-growth, closed-canopy forest.

highgrading. Removal of the best or most desired of a species from a renewable resource. Highgrading occurs in fisheries where smaller and less-valued fish may be dumped at sea so as to maximize the hold capacity with more valuable fish. In forestry, highgrading involves the removal of the best trees leaving behind low-quality stands of trees.

high-level radioactive waste. See *radioactive waste*.

high-sulfur coal. Coal which contains a high proportion of sulfur compounds that can be released as emissions when burnt. Sulfur concentrations vary substantially across coal deposits.

high yielding varieties (HYVs). Domesticated plant varieties that, with the appropriate application of fertilizers, can provide higher yields than traditional varieties. See *green revolution*.

HIV. See *human immunodeficiency virus*.

Holocene. The current geological epoch which began at the end of the last ice age 10–12,000 years ago.

homeostasis. The tendency of a system to maintain an equilibrium through negative feedbacks.

hominids. A family of primates called Hominidae from which modern humans (*Homo sapiens*) are the last remaining species. Modern humans are believed to have evolved in Africa some 100,000 years ago.

homogeneous function. A set of functions widely used in economic analysis that have the following property where if f(x) is the function then

$$f(ax) = a^k f(x)$$

for all values of a greater than zero, and where k is defined as the degree of homogeneity.

Homo sapiens. See *hominids*.

homoscedasticity. A term used in statistical models to refer to residuals or error terms that have equal variance around a mean of zero. See *heteroscedasticity*.

horizontal equity. Outcomes in which similarly situated people are treated equally (for example, people with the same income are taxed at the same rate). See *vertical equity*.

horizontal integration. Economic term for the agglomeration of companies at the same level of production, such as a merger of oil companies.

hormone. An organic compound used to control various functions in plants and animals, including growth and sexual development.

hormone disrupter. See *endocrine disruptors*.

host. Organisms that sustain parasites. See *parasites*.

hot air. Term used when referring to Annex I countries of the Framework Convention on Climate Change (or Annex B countries in the Kyoto Protocol) whose emission obligations far exceed their current level of emissions. Should trading in greenhouse gases (GHG) emissions be permitted, such countries (for example, Russia) could sell their "hot air" to other countries who could then use them to meet their GHG obligations without reducing world GHG emissions.

Hotelling model.   Model that explains the spatial location of firms and suggests that firms in a similar business are likely to locate close to each other.

Hotelling rent.   Named after the economist Harold Hotelling, this is a method of valuing non-renewable resources defined as the marginal net return on the last unit of the resource extracted.

Hotelling's lemma.   A result that states that the output supply and input demand functions can be derived from a profit function by taking its partial derivative with respect to the output price and input prices.

Hotelling's rule.   Named after the economist Harold Hotelling, it is a rate of extraction of a non-renewable resource such that the net price of the extracted resource rises at a rate equal to the rate of interest in an economy. Hotelling's rule is an equilibrium condition and will hold true only under special conditions such as perfect foresight and information, and competitive markets.

human capital.   The learned and inherent abilities that are embodied in individuals.

human genome project.   An international project involving thousands of researchers that has mapped the human genome.

human immunodeficiency virus (HIV).   The virus responsible for AIDS. See *acquired immune deficiency syndrome.*

Humboldt current.   Also known as the Peru current, it is a cold ocean current that travels north along the western coast of South America. Upwelling of nutrients from the current provides sustenance for an abundance of marine life and one of the world's largest fisheries, the Peruvian anchoveta.

humidity.   A measure of the amount of water in a given volume of air. The higher the temperature, in general, the more water the air can hold.

humus.   Partially decomposed plant and animal material found in soils. It is important for retaining moisture and maintaining soil fertility.

hunter-gatherer.   Term used to describe a pre-agricultural mode of living whereby animals are hunted, and plants and fruits are gathered to provide sustenance.

hybrid. Term used to describe the progeny or offspring of a union between different breeds or species.

hydrocarbons. Compounds of hydrogen and carbon formed from organic material, including fossil fuels. Their combustion contributes to the emission of greenhouse gases, such as carbon dioxide and methane.

hydrochlorofluorocarbons (HCFCs). A widely used substitute for chlorofluorocarbons (CFCs) used to decrease damage to stratospheric ozone. For equivalent weight, HCFCs are at least 10 times less damaging to stratospheric ozone than CFCs.

hydro-electric. Electrical power that is generated from water passing through turbines. To ensure a reliable and steady supply of electricity, dams are built as reservoirs for periods of low water flow.

hydrofluorocarbons (HFCs). Organic carbon compounds that have been used to replace chlorofluorocarbons (CFCs) and hydrochlorofluorocarbons (HCFCs) in many uses, especially in refrigerants. The global warming potential of HFCs can be over a thousand times greater than carbon dioxide and thus these are included as greenhouse gas emitters that should be controlled by Annex I (Annex B countries in the Kyoto Protocol) countries.

hydrogen oxides. Naturally occurring compounds that contribute to stratospheric ozone depletion.

hydrogen sulfide (H_2S). A poisonous and flammable gas that smells like rotten eggs and is often produced in oil production and in oil refineries.

hydrological cycle. The physical processes by which water is recycled, whereby precipitation on land is followed by infiltration of water tables and run-off, transpiration from plants, and evaporation to complete the cycle. At sea, precipitation is followed by evaporation.

hydrology. The science and study of the hydrological cycle, groundwater and water management.

hydrophyte. Plant adapted to growing in water or in very wet environments, such as a water lily.

hydroponics.   The practice of growing plants in aquatic media under controlled conditions.

hydropower.   The production of electricity from turbines propelled by moving water from either in-stream flows or the diversion of dammed waters through channels. See *hydro-electric*.

hydrothermal vents. Heated water from geothermal activity that emerges in the deep ocean.

hyperdepletion.   A relationship between catch per unit of effort (CPUE) in a fishery and total abundance where the CPUE drops much more rapidly than does overall abundance. See *hyperstability*.

FURTHER READING
Hilborn and Walters (1992).

hyperstability.   A relationship between catch per unit of effort (CPUE) in a fishery and total abundance where the CPUE drops much less rapidly than does overall abundance. See *hyperdepletion*.

FURTHER READING
Hilborn and Walters (1992).

hyperthermophiles.   Bacteria that live at extremely high temperature environments that range between 80 degrees Celsius and 150 degrees Celsius. Some scientists have hypothesized that hyperthermophiles may have been some of the earliest life on our planet.

FURTHER READING
Davis (1999).

hypolimnion.   A cooler and lower layer of water in a lake.

hypothesis.   A statement used to explain a phenomenon. A hypothesis should be able to be tested and falsified. See *null hypothesis*.

hypothetical bias.   The so-called bias that arises when persons are asked a hypothetical question in which they have limited ability to provide a meaningful response.

hypothetical resources.   Deposits that have not yet been discovered but that can reasonably be expected to be discovered based on a similar geological structure to that in which current deposits are found.

I

ice age. A period when the quantity of snow and ice on the earth's surface was at least double what it is today. The earth has experienced many ice ages going back more than 570 million years. The most recent ice age ended about 10–12,000 years ago and, given that there have been at least 20 interglacial (between ice age) periods in the past 3 million years, the current warm period in which we live is viewed by some scientists as an interglacial rather than post-glacial period.

FURTHER READING
Wilson, Drury and Chapman (2000).

ice-core samples. Samples of ice obtained from glaciers that can be used to determine the composition of the earth's atmosphere at the time the glacier was formed.

icehouse effect. Term used to describe the factors that can combine to produce an ice age.

identity matrix. A square matrix whose elements along its main diagonal (elements a_{ij} where $i=j$) are all equal to one and every other element equals zero.

IGOs. Acronym for intergovernmental organizations, such as the World Trade Organization (WTO).

illite. A type of clay formed by the weathering of feldspars and mica.

imission. Term used in Germany to describe emissions from distant sources that have negative local effects.

impact analysis. An analysis where decision makers weigh the consequences of different actions or policies, as measured by changes in variables stemming from the different actions.

impact area/zone. The area in which the environment is likely to be affected by a specific policy.

implicit costs. Costs that are not directly considered by firms or consumers but are, nevertheless, a cost of production imposed on either the firm itself or others. See *explicit costs*.

import quotas.   Legal limits placed on the quantity of a good that can be imported into a country.

improvement cutting.   The removal of (economically) undesirable species in order to promote forest health and growth.

in-breed.   To reproduce from members of a species that are very similar genetically.

in camera.   Latin term meaning "in private".

incentive compatibility.   Term used to describe a situation where a personal or individual interest coincides with the intentions of a regulator or the public interest. For example, the deductible in an insurance contract provides the person buying the insurance with an added incentive to prevent the loss.

inch.   Unit of length equal to 2.54 centimeters.

incidental taking.   Under the Endangered Species Act in the USA, an incidental taking is defined as when a landowner is engaged in otherwise lawful activities but indirectly harms an endangered species.   See *taking*.

income effect.   The change in consumption that results from a change in the price of good in terms of its effect on the real income of an individual. See *substitution effect*.

income elasticity.   The change in the consumption of a good or service following a change in income. If the change in demand is in the same (opposite) direction as the change in income, the good in question has a positive (negative) income elasticity.

increasing returns to scale.   A situation where if all the inputs in a production process are increased by the same percentage, the output will increase by a higher percentage.   Increasing returns to scale tend to lead to large-scale plants, as the greater the amount produced the lower the per unit cost. See *decreasing returns to scale* and *constant returns to scale.*

incrementalism.   Making gradual rather than abrupt changes in policy.

independence of irrelevant alternatives.   A concept used in welfare economics which states that the introduction of an alternative choice should not alter the relative ranking of two other choices. Also used as a criterion to determine the outcomes of certain games in game theory.

independent variable.   Variable in a regression that is set exogenously from the model and helps determine the dependent variable.

indexation.   Automatic increases in wages, the value of pensions and other benefits that follow increases in the rate of inflation.

indicated resources.   Deposits that have been discovered and the size of which estimated based on sampling and projections derived from similar types of deposits.

indicator species.   A species that can be used to asses the quality of the environment. Indicator species often have a low level of tolerance for a particular aspect of the environment and thus are most vulnerable to changes in ecosystems.

indifference curve.   A graphical representation of the different combinations of goods that yield the same level of utility for an individual.

indigenous knowledge.   The body of knowledge of local inhabitants regarding the physical functions, uses, and interdependence of resources and living organisms found within their environment. See also *traditional knowledge.*

indigenous species.   A plant or animal species native to a particular region or area.

indirect objective function.   See *optimal value function.*

indirect taxes.   Taxes imposed on the sale of goods and services rather than income. See *direct taxes.*

indirect utility function.   A function that provides a measure of the utility of an individual, but is a function of prices and income rather than a function of the quantities of goods and services consumed.

indirect valuation.   The measurement of the impact from a change in pollution on some aspect of the environment (such as an increase in air pollution on human health or on plant growth).

Individual Fishing Quotas (IFQs).   Term   used   for   individual transferable quotas in the US fisheries. See *Individual Transferable Quotas*.

individual rights.   Property rights to an asset or resource held by an individual. See *community rights*.

Individual Transferable Quotas (ITQs).   An output or quantitative management tool for regulating fisheries. The total allowable catch is allocated among fishers in the form of harvesting rights that can be traded.   Provided that the ITQs are viewed as an exclusive and durable property right, they can help overcome "the race to fish" associated with input-based controls.

FURTHER READING
Grafton (1996).

Individual Vessel Quotas (IVQs).   Term   used   for   individual transferable quotas in some Canadian fisheries. See *Individual Transferable Quotas*.

induction.   See *inductive method*.

inductive method.   A method of reasoning by which facts and general observations are used to form a generalization or a theory or a scientific principle. See *deductive method*.

industrial revolution.   Term used to describe the rapid transformation of an essentially agriculture-based economy to one where most of the output and employment is generated by the industrial sector. The first industrial revolution began in the United Kingdom at the beginning of the nineteenth century and coincided with a massive increase in the use of natural resources, a social transformation and a major change in the environment and landscape.

industry concentration.   The share of a market served by the largest firms. The larger the market share for a given number of firms, the greater is the industry concentration. See *concentration ratio*.

inelastic.   Term used to describe the consumer response or producer response to a change in price. A good is inelastic if the percentage change in demand (or supply) is less than the percentage change in the price. For example, if the price of a good fell by one percent and demand increased by less than one percent, the good would have an inelastic demand at that price. See *elastic*.

infant industry argument. The notion that to enable an industry to be internationally competitive it may first be necessary to protect its domestic market through import quota tariffs.

infant mortality rate. The annual number of deaths of children under the age of one year per thousand live births in a population.

inferior good. A good or service which has a demand that is inversely related to a consumer's income.

inferred resources. Deposits that may exist contiguous to existing deposits in unexplored areas.

infiltration. The absorption of water by soil.

inflation rate. A measure of the change in the price level of a defined set of goods and services over time. See *consumer price index*.

infra. Latin term meaning "below".

infrared radiation. Long-wave radiation that is emitted from the earth's surface and is effectively absorbed by water vapor and greenhouse gases. See *greenhouse gases*.

infrastructure. The physical communication, delivery and transport-ation systems within an economy.

initial conditions. The starting values of stocks or state variables in a dynamic model.

inorganic. Substances that either have no carbon, or only carbon combined with elements other than hydrogen.

in perpetuum. Latin term meaning "forever".

in personam. Latin term meaning "in person".

input controls. A type of management that seeks to control the inputs used by resource users so as to control the total effort and/or harvest from common-pool resources. Input controls are widely employed in fisheries, but have frequently failed to prevent excess fishing effort or capital stuffing because fishers are often much better at circumventing input controls than are regulators at devising such regulations. See *output controls.*

input–output analysis.   A method of modeling an economy by specifying the relationship between inputs, intermediate outputs, outputs and final demands in an economy. The methodology can be used to examine the impact on the rest of the economy of a shock in a particular sector.

inputs.   Factors of production used to produce intermediate or final outputs. See *factors of production* and *outputs.*

in situ.   1. Latin term for the natural state.   2. Attributes and characteristics of a natural resource or environmental asset that arise from its current state.   For example, a growing forest provides in-situ value in terms of carbon sequestration.

in situ price.   See *in-situ.*

insolation.   The amount of solar radiation that reaches the surface of an object in a defined period of time.

institutional economics.   A sub-discipline of economics that emphasizes the importance of institutions and social structures and their affect on economic performance.

FURTHER READING
Eggertsson (1990).

institutions.   The anthropogenic rules and norms of behavior that govern all human interactions.

in-stream augmentation.   Supplementing the flow of water within a stream through either increased supply of water to the stream, or practices that increase flows (such as releasing more water from a dam during periods of low stream flow).

in-stream flow protection.   The maintenance of water flows within a stream, which may be diminished through the diversion of water for irrigation purposes, or for power generation.

in-stream flows.   The state in which water remains in its natural course (e.g., streams, creeks, or rivers) as opposed to water that has been diverted artificially for other purposes (e.g., irrigation, reservoirs, drinking, etc.).

in-stream rights.   The right to enjoy a minimum amount of water flow within a stream or river.

in-stream use.   Use of water flowing in a stream or river, such as for fishing or navigation.

insulation.   Method by which heat is retained in a body by reducing conduction. Insulation of houses and offices can result in significant savings in terms of energy use.

intangible benefits.   Benefits that accrue to individuals that cannot be quantified in financial terms.

integrated pest management (IPM).   An ecological approach to the management of pests which eschews the use of pesticides. IPM involves a variety of techniques, including biological controls and the planting of mixed crops to reduce, but not necessarily eradicate, pests on crops.

integrated pollution prevention and control.A broad-based ap-proach to the management and control of pollution that considers all potential sources and their possible impacts across the environment.

integrated solid waste management.   A combination of recycling, composting, incineration and judicious use of landfills to reduce the amount of solid waste produced by society.

integratibility.   In utility or demand theory, the requirement that the Slutsky matrix of substitution terms be symmetric and semi-definite. Integratibility ensures that preferences can be recovered by integrating a system of demand functions to recover the expenditure function, which in turn can be used to derive the indirect and direct utility functions.

intellectual property rights (IPRs).   Patents or copyrights that provide the originators of creations, innovations or ideas with a legally enforceable right to exclude others from appropriating their intellectual capital without their prior consent. IPRs exist for a range of intellectual creations that include the written word, inventions, art, techniques, image, and designs but such protection is, in general, not provided for indigenous or traditional knowledge. See *property rights* and *indigenous knowledge*.

intensive agriculture.   Agricultural practices associated with high yields but with concomitant use of high inputs of pesticides, herbicides and artificial fertilizers. Intensive agriculture in terms of livestock often involves the production of large quantities of excreta that can impose a major burden on the environment and can lead to contamination of soil and water.

intensive forest management.   A range of forest practices such as pruning, fertilizing and spacing designed to yield higher timber quality, increased harvest volumes, or reduced rotation periods.

intensive margin.   Term used for a plot of land or stand of timber where the value of the marginal product associated with a strip of the land, or of a tree in the stand, exactly equals the marginal cost of production.

FURTHER READING
Van Kooten and Bulte (2000).

inter alia.   Latin term meaning "among other things".

Inter-American Development Bank (IADB).   Bank established in 1959 by member countries that is mandated to further the economic and social development of nations in the Caribbean and Latin America.

inter-basin transfer.   The transfer of water across watersheds for a particular purpose. Inter-basin transfers, widely practiced in the Western United States, have been highly controversial because of their deleterious effect on ecosystems.

intercropping.   The practice of growing different crops between the rows of another.

intergenerational equity.   The notion that the present generation should consider the needs of future generations in current decision making. Intergenerational equity is a fundamental feature of sustainable development. See *equity* and *sustainable development.*

interglacial period. A period of time in between two ice ages. The earth has experienced at least 20 interglacial periods in the past 3 million years. See *ice age.*

Intergovernmental Panel on Climate Change (IPCC). A panel which involves thousands of scientists from around the world that is funded by the governments of developed countries to undertake research on climate change.

intermediate good. Goods that are used in the production process to produce other goods. For example, wheat is an intermediate good as it is the raw material for bread and other wheat products.

internal rate of return (IRR). The discount rate in cost-benefit analysis at which the present value of benefits equal the present value of costs. See *cost–benefit analysis.*

internalize. Term given for the process by which individuals or firms explicitly consider the costs or benefits that they impose on others from their actions or decisions.

International Bank for Reconstruction and Development (IBRD). More commonly known as the World Bank. It was established in 1944 and its mandate is to lend to member countries to assist them in their social and economic development.

International Council for the Exploration of the Sea (ICES). An international body that is almost 100 years old. Its 19 member countries promote and coordinate fisheries and marine research.

International Energy Agency (IEA). An agency formed in 1973 by developed countries in response to the first oil price shocks. The IEA provides information on global energy production and consumption.

International Joint Commission (IJC). A commission and advisory body formed in 1909 by the USA and Canada to oversee the management of their boundary waters. Since the early 1970s, the IJC has been instrumental in setting pollution guidelines for the Great Lakes and has helped initiate considerable improvements in the quality of the water in the lower Great Lakes, especially Lake Erie.

International Monetary Fund (IMF).   An international organization formed by the Bretton Woods Agreement in 1944 that now has over 180 members. Its membership finances special drawing rights that enable countries to deal with temporary balance of payments difficulties. In providing funding to countries in need, the IMF may also specify conditions in terms of fiscal and monetary policy that must be followed before assistance is provided.

International Organization for Standardization (ISO).   An international organization based in Geneva that is mandated to provide international standards. See *ISO 14001.*

International Union for the Conservation of Nature (IUCN).   Also known as the World Conservation Union.   It was formed in 1948 by governments and non-governmental organizations and today has over 139 countries represented in its membership. Its mandate is to conserve the integrity and diversity of nature in an equitable and ecologically sustainable way.

International Whaling Convention (IWC).   Convention signed in 1946 that established the International Whaling Commission, mandated to preserve whale stocks. The Commission has been highly controversial and failed to prevent the near extinction of several whale species. More recently, a moratorium on commercial whaling has helped some species to increase in number but controversy still remains as some members of the commission would like to ban all commercial whaling in perpetuity while other countries would like to resume commercial whaling.

interstadial period.   A period of less than 2,000 years of warmer temperatures, but which occurs within a period of glaciation. See *interglacial period.*

intertemporal efficiency.   See *dynamic efficiency.*

intertidal community.   A community of marine organisms living in coastal areas and along a shoreline that is submerged during high tides and exposed during low tides.

intertropical convergence zone (ITCZ).   A low pressure zone which changes with the seasons but is always found close to the equator where the rising of warm tropical air caused by surface heating creates an area of low pressure. The variability of the ITCZ causes seasonal droughts in Africa and Asia.

interviewer bias.   The bias in responses by respondents in a survey due to the particular behavior of an interviewer.

intrinsic value.   Value of good, asset or aspect of the environment that is separate from its value in use. See *non-use value.*

FURTHER READING
Fisher and Raucher (1984).

inverse demand function.   A relationship between the quantity of a good consumed and the marginal willingness to pay for the last unit of the good consumed, holding everything else constant.

inversion.   Shortened term for a temperature inversion where air temperature is either unchanged or increases with altitude. The effect is that air pollution can be trapped by warmer air above, as occurs in Los Angeles at certain times of the year.

invertebrates.   Animals without a backbone that include insects and crustaceans.

investment.   The expenditures associated with the purchase of new capital goods. See *capital good.*

invisible hand.   The notion, first described by Adam Smith, that individuals pursuing their own self-interest in their particular trade or profession can provide benefits for society as a whole by ensuring goods and services are supplied that are demanded.

in vitro.   Latin term meaning "existing outside of a living creature".

in vitro fertilization.   Fertilization of an egg outside of a living organism.

in vivo.   Latin term meaning "existing within a living creature".

involuntary risk.   Risk imposed on individuals without their consent, such as the effects of air pollution.

ion.   An atom which has lost or gained an electron and is either negatively or positively charged.

ionization.   The process by which atoms become ions.

ionizing radiation. Radiation capable of causing atoms to become negatively or positively charged by the gain or loss of electrons. Ionizing radiation can be caused by various forms of electromagnetic radiation and can lead to various health problems, including cancers.

ionosphere. The outermost layer of the earth's atmosphere.

IPRs. See *intellectual property rights.*

ipso facto. Latin term meaning "by the fact itself".

irradiation. Exposure to ionizing radiation. Irradiation can be used to protect food from contamination by micro-organisms but is controversial because of concerns about its possible health effects.

irreversibility. The notion that many processes and developments are difficult, or may even be impossible, to change. For example, the extinction of a species is irrevocable.

island biogeography. The study of the positive relationship between habitat size, remoteness and the number of species found within the habitat.

IS–LM model. A macroeconomic model developed by Sir John Hicks and based upon John Maynard Keynes' book, *General Theory of Employment.* The model includes IS (investment-savings) that represents the combinations of interest rates and savings that generate equilibrium in investment and savings and LM (liquidity-money supply) that ensures that money demand equals money supply for different combinations of interest rates and income.

ISO 14001. A set of environmental standards developed by the International Organization for Standardization that assists organizations to improve their environmental management.

isocost. Combinations of input quantities for given input prices that generate the same total cost of production.

isohyet. Line on a map joining areas of equal average rainfall.

isoquant. The combinations of input quantities that produce the same level of output.

isotope.   Forms of an element that differ in terms of the number of neutrons. Isotopic forms of oxygen and carbon have been used for the dating of past events.

Itai-Itai.   A degenerative bone disease caused by excessive amounts of cadmium in the body.

IUCN.   See *International Union for the Conservation of Nature.*

ius disponendi.   Latin term meaning "the right to sell or transfer". See *transferability.*

ius excludendi.   Latin term meaning "the right to prevent interference". See *exclusivity.*

ius fruendi or usufructus.   Latin term meaning "the right to withdraw". See *withdrawal right.*

ius possidendi.   Latin term meaning "the right to possess or own". See *quality of title.*

ius utile.   Latin term meaning "the right to enjoy".

J

Jacobian matrix.   A matrix consisting of first-order partial derivatives that is used to test for functional dependence among a system of equations.

FURTHER READING
Grafton and Sargent (1997).

Jensen's inequality.   An inequality where the expected value of a concave function of a random variable, $E[g(x)]$, is equal to or less than the concave function of the expected value of the random variable, $g(E[x])$, if $g(.)$ is a concave function, x is the random variable and $E[]$ is the expectations operator. See *expected value* and *random variable*.

FURTHER READING
Greene (1997).

jet stream.   A stream of air found at the upper boundary of the troposphere (8 to 15 kilometers above the earth's surface depending upon the latitude) which can travel as fast as 300 kilometers an hour.

joint and several liability.   Term used to describe a legal standard whereby a single person or party is liable for all damages despite the fact that the party may have been responsible for only a part of the total damages.

joint implementation.   A term defined under the 1997 Kyoto Protocol that allows Annex B countries, which have obligations to reduce their greenhouse gases under the Framework Convention on Climate Change, to mitigate greenhouse gas emissions or enhance sinks in Annex B countries by cooperation and investments. The reduction in emissions from the cooperation and investments may then be used as emissions reduction units to help meet national greenhouse gas reductions required under the Protocol. See *emission reduction credit*, *Kyoto Protocol* and *clean development mechanism*.

jointness in inputs.   A production relationship where all inputs are required to produce all outputs.

joint product.   Goods or services that are produced jointly in more or less fixed proportions, such as mutton and wool.

joule. A unit of energy defined as the work from the force of one newton exerted over one meter.

Jurassic. A geological time period that occurred from 208 to 144 million years ago.

just compensation. The fair value to be paid to an individual or group in compensation for the environmental degradation or harm created by the development or use of the resource by others.

K

Kalahari Sands. Aeolian sands of Quartenary origin that are found from Southern Africa north into the Republic of Congo.

Kaldor Potential Compensation. One social state is preferred to another if there exists another third state that could be reached after making the move to that state, and that third state is Pareto superior to the initial state (the potential state that could be reached is preferred by all parties to the initial state). There is no assurance, however, that this third state will be reached and this criterion can lead to contradictory results.

kaolinite. A clay derived from the weathering of feldspar-rich rocks, usually under acidic conditions in humid tropical soils.

katabatic wind. A downward air-flow in a valley caused by cooling at higher altitudes at night-time.

kelp. Large and often fast-growing marine plants that serve as important habitat for marine organisms.

Kelvin scale. A temperature scale named after the British scientist Lord Kelvin. Each degree Kelvin is equivalent to one degree on the Celsius scale but the scale starts at 0 at absolute zero or –273 degrees Celsius.

keystone species. A dominant species in an environment such that its removal has an important impact on all other species.

kilogram. Metric measure of mass that equals 2.204 pounds in the imperial measure.

kilometer. Metric measure of distance that equals 0.621 of a mile.

kinetic energy. The energy of an object in motion. It depends on its mass and velocity.

kingdom. Broadest classification of organisms that includes the animal, plant, fungi, monera and protista kingdoms.

knot. One nautical mile per hour.

known deposit.   See *proved reserves*.

Krakatoa.   An Indonesian volcanic island that erupted in 1883, depositing enough material in the atmosphere to affect world temperatures in 1884.

Krebs cycle.   The chemical reactions in cells that provide the energy organisms use to function.

k-strategists.   Term used to describe animals that spend a great deal of time and energy in rearing their young, such as primates.

Kuhn–Tucker conditions.   Set of conditions for nonlinear programming problems that, if satisfied, are the necessary conditions for an optimum.

kurtosis.   A measure of how a unimodal probability distribution differs from a normal distribution in terms of whether it is flatter or more pointed.

Kuznets curve.   A supposed relationship observed by Simon Kuznets between per capita income and the degree of inequality in an economy. According to Kuznets, inequality rises with economic development and then peaks, after which point it declines, even as growth continues. See *environmental Kuznets curve*.

FURTHER READING
Ray (1998).

kwashiorkor.   Human disease associated with a low–protein diet, common in children in many developing countries.

Kyansur forest disease.   A viral disease transmitted by ticks and found in southern India. Deforestation has increased the incidence of the disease as livestock near the forest provide ready hosts for the ticks.

FURTHER READING
Grifo (1999).

Kyoto mechanisms.   Mechanisms designed to reduce the overall cost of Annex I (Annex B countries in the Protocol) meeting their greenhouse gases (GHG) emission obligations.

Kyoto Protocol. The 1997 agreement by parties to the Framework Convention on Climate Change that sets binding emission constraints on Annex I countries (listed as Annex B countries in the Protocol) for the period 2008–12. Collectively, Annex B countries are committed to a 5 percent reduction in greenhouse gas emissions by 2008–12 relative to base levels in 1990. Some Annex B countries (such as Australia) secured increases in emissions over 1990 levels, some agreed to meet 1990 levels (such as New Zealand) and some agreed to exceed the 5 percent reduction (such as Germany). The Protocol allows for the possibility of carbon sinks and emissions trading although the precise rules are to be determined upon at a later date.

L

labor-intensive. Goods and services that use more labor, relative to other factors, in their production process.

labor–leisure trade-off. The supposed trade-off that people face when determining how many hours they should work. Greater hours worked means higher income, but the trade-off is reduced leisure hours.

labor productivity. The real value of output per unit of labor input, usually measured per hour or person days.

labor theory of value. The notion, attributed to Karl Marx, that the measure for the value of goods and services should be the amount of labor (such as person hours) devoted to its production.

Labrador current. A cold current that flows southwest from Greenland towards eastern Canada, and the coasts of Labrador and Newfoundland.

Lagrangian function. Function formed by the objective function from an optimization problem plus each of the constraints each multiplied by a Lagrangian multiplier. See *Lagrangian multiplier.*

FURTHER READING
Grafton and Sargent (1997).

Lagrangian multiplier. A variable created to help solve optimization problems with constraints. It is treated like a choice variable in the optimization problem. See *Lagrangian function.*

lag time. The period of time from an initial event before its effects can be observed or realized.

lahars. Landslides of volcanic origin.

Lake Nyos. Site in Cameroon where in 1986 a release of carbon dioxide from the lake asphyxiated almost two thousand people.

Lamarckian evolution. The discredited theory, named after Jean Baptiste Lamarck, that evolution occurs due to the inheritance of acquired characteristics. For example, Lamarck would have believed that a person with an acquired characteristic, such as a darker skin because of a suntan, could pass this characteristic on to her offspring.

land grant. Land given by the government to individuals, companies, or institutions.

land rent. The payment associated with a site that can be attributed to some scarce factor about the land, such as fertility or location, and which can be taxed away without inducing a change in how the land is used.

land-use planning. The practice of managing land for the most desirable purposes. Such planning ensures that green spaces exist in urban environments and prevents undesirable industries from locating in residential neighborhoods.

La Niña. An irregular event where the waters off the coast of Chile and Peru are unusually cold, in contrast to El Niño events where the waters are unusually warm. The last La Niña event peaked in December 1998. See *El Niño*.

FURTHER READING
Allaby (1996).

Laspeyres index. A commonly used index that employs weights from a base period. For example, a Laspeyres price index may be defined as p_1q_0/p_0q_0 where q_0 is the quantity consumed in the base period, p_0 is the price of the good in the base period and p_1 is the price of the good in the current period. See *Paasche index*.

laterite. A tropical soil that contains high levels of iron and/or aluminum oxides, and is therefore of poor quality.

laterization. Process by which laterite soils are formed. It can be accelerated by anthropogenic influences, such as deforestation. See *laterite*.

Laurasia. A supercontinent that broke from the even larger super-continent Pangea and eventually split and formed into North America (including Greenland), Asia, and Europe. See *Pangea*.

lava. Magma that reaches the earth's surface.

law of conservation of energy. The physical law that energy can neither be created nor destroyed but can only be transformed or transferred to another place.

law of demand. The tendency for most goods and services that the quantity demanded will increase as the price decreases, all things else being equal.

law of large numbers. The notion that as the number of observations becomes very large, the incidence of a particular event (as a ratio of all observations) converges to the probability of such an event occurring.

law of supply. The tendency for most goods and services that the quantity supplied will increase as the price increases, all things else being equal.

LDC. Acronym for Less Developed Country.   See *developing countries*.

leachates. Soluble compounds dissolved and carried by water percolating through the ground from any anthropogenic sub-surface structure, such as landfills or tailings.

leaching. The transportation of soluble material (organic and inorganic) from soils or waste sites through the percolation of water. Agricultural run-off may include fertilizers and pesticides that have leached from the soil.

lead. A widely used metal in batteries and, in some countries, still used as an additive to paints and fossil fuels. Lead readily accumulates in the body and chronic (low-level) exposure may affect the development of children (especially cognitive abilities) and, at high levels of exposure, may even result in death.

lead poisoning. See *lead*.

least squares. A method of estimating parameters in a regression model by minimizing the differences between the predicted values from the model and the observed values from the data, multiplied by some weight. If the weights are identical and equal to one, the method of estimation is called ordinary least squares. See *ordinary least squares*.

Le Chatelier principle. 1. The notion that a physical system in equilibrium responds to a shock in a way that minimizes its effects. 2. In economics, the notion that the economic response to changes in prices is greater in the long run when all variables can change than in the short run when some variables may be fixed.

legume. Plants of the pea family (including beans, lentils, etc.) that have the ability to "fix" nitrogen and enrich the soil because of the nitrogen-fixing bacteria that grow on their roots.

leishmaniasis. A debilitating and sometimes fatal disease spread by sandflies and caused by protozoa. The incidence of the disease is increased with deforestation and the settlement of equatorial forests that increase human exposure to sandflies that have bitten animals with the disease.

FURTHER READING
Grifo (1999).

lentic. Term used for standing or stagnant freshwater.

lex non scripta. Latin term for the common law.

lex scripta. Latin term for written or statute law.

liability. A legal rule that makes it possible for individuals, firms or even governments to be held legally responsible for their actions.

libertarianism. Term used to describe the notion that individual liberties are paramount and that the state should not interfere or infringe upon these rights. In its extreme, some libertarians argue that taxation contravenes the rights of individuals to spend the money they earn as they see fit.

lichen. An association of fungi with green algae.

life-cycle analysis. A scientific accounting of material and the effects of throughput of a natural resource from initial extraction, through production and consumption, and finally to the ultimate emission of residuals to the environment.

life-cycle model. A method of valuing risk that models the way a change in external factors causes a change in an individual's probability of dying.

FURTHER READING
Cropper and Sussman (1990).

life expectancy.   The age that half the population is expected to live to or beyond.

life table.   A mortality table that specifies the average remaining years of life at each age for a defined population.

light green technologies.   Processes that may benefit the environment, such as the development of fuel-efficient vehicles, but that were not necessarily developed to reduce environmental degradation. See *dark green technologies.*

lignite.   A type of coal with properties in-between that of peat and bituminous coal.

likelihood function.   Function defined as
$$L(x, \theta) = f(x_1, \theta) \cdot f(x_2, \theta) \ldots f(x_n, \theta)$$
where x_1, x_2, ... x_n are observations of a random variable x and the estimated parameters θ are maximum-likelihood estimators that maximize the likelihood function (or a logarithmic transformation of the function). See *maximum likelihood.*

FURTHER READING
Ramanathan (1992).

likelihood ratio test (LR).   A widely used statistical test used to evaluate hypotheses. The test compares the values of the log-likelihood function with ($\ln_r$) and without ($\ln_{ur}$) restrictions where
$$LR = -2\{\ln_r - \ln_{ur}\}$$
is distributed as a chi-squared statistic. See *likelihood function.*

limit.   The value of a function as the argument of the function approaches a particular value.

limited dependent variable.   Econometric models in which the dependent variable can only be observed over a particular range due to either a lack of observations, or the fact that the data only exists for a particular subset.

limited entry.   Term used to describe controls placed on the number of resource users, and possibly their inputs, in the use of natural resources. Limited entry is a common form of regulation in fisheries. See *input controls.*

limited-user open access.   A property-rights regime where the number of resource users is regulated, but no other restrictions exist in terms of how the resource is exploited.

limiting factor.   A factor, element or resource in the environment that may restrict the growth of individuals or distribution of a species or communities. In deserts, water is often a limiting factor that prevents their colonization by plants and animals that are not adapted to water-scarce environments.

limits to growth.   See *Club of Rome*.

Lindahl equilibrium.   A Pareto-efficient equilibrium involving both private and public goods where individuals agree to the same level of provision of public goods for a set of lump-sum transfers of income. See *Pareto efficiency* and *public good*.

FURTHER READING
Cornes and Sandler (1996).

linear function.   A function where the variables are not raised to any power except one and are not cross-multiplied by any other variable. For example, $y = 3a$, is a linear function but $y = 3a^2$ is not.

Linear growth.   Growth in a variable over time such that it increases by the same fixed amount each time period.

linear probability model.   A regression model in which the dependent variable is either 1 (representing the occurrence of an event) or 0 (representing the absence of an event).

linear programming.   A widely used mathematical programming technique for problems that are linear in the objective function and in the constraints.

FURTHER READING
Grafton and Sargent (1997).

lipids.   Substances that are soluble in fats but not water.

liquefied petroleum gas (LPG).   Term used for propane or butane, or a mixture of the two gases, that can be stored liquefied under pressure.

liquidity.   Term used to describe how quickly an asset can be converted into cash.

liquidity preference.   Term used to describe the demand, or the desire, to hold cash over other assets.

listeria monocytogenes.   A pathogen found in the feces of some animals that is responsible for widespread food contamination as it can be transferred via water, soil and animal tissues. Although most people can tolerate the pathogen, it can be particularly harmful to pregnant women and can cause miscarriages.

FURTHER READING
Fox (1998).

lithology.   The science and study of the structure of minerals and rocks.

lithosphere.   The subsoil layers of the earth including its core, mantle and crust.

litter fall.   The accumulation of organic debris on the forest floor from falling leaves and other material.

little ice age.   A cooling trend in the earth's temperature, especially in the northern hemisphere, that spanned the period from the sixteenth century up to the early or mid-nineteenth century.

littoral zone.   The ecological zone between the low and high water marks of lakes, rivers and the sea.

load factor.   The ratio of the average electricity demand to the maximum peak load that the electrical system can support.

loam.   Soil that is a mix of clay, silt and sand.

locally unwanted land use (LULU).   A land use that would decrease the utility of local constituents if sited "nearby".

local pollution.   Pollution from which most or all the degrading effects to human, environmental, and/or ecological well-being occur locally.

local public good.   A public good the benefits fo which are limited to a small area or a particular population, such as a city park. See *public good*.

FURTHER READING
Cornes and Sandler (1996).

location theory. A set of theories used to explain land use and location of economic activities. German writers, such as Von Thunen and Weber, were the pioneers in location theory. Von Thunen provided a theory as to why farmers produce the crops they do in relation to their markets, and Weber provided a theory about where firms locate their production relative to their raw materials and market.

loess. Silt deposited by wind which originates from glacial plains or desert regions.

logarithm. The power or exponent to which a base must be raised to obtain a particular number. For example, using the base 10, ln 100 = 2 since $10^2 = 100$.

logarithmic scale. A scale that uses a base equal to 10 such that an increase in the scale by one represents a 10 fold increase in the value of the variable being measured. For example, the Richter scale is logarithmic such that an earthquake of 7.0 generates a tenfold increase in the intensity of seismic waves than an earthquake of 6.0. See *logarithm*.

logistic growth. Natural population growth characterized by a logistic growth function and that is characterized by density dependence. A generalized density-dependent growth function is defined by

$$ f(x) = rx\left(1 - \frac{x}{K}\right)^{\alpha} $$

where $f(x)$ is growth in the population, x is the population, r is an intrinsic growth rate, K is the carrying capacity and α is a parameter. If α is less (greater) than unity the growth function is skewed to the right (left). As x approaches the carrying capacity the population growth approaches zero. See *carrying capacity*.

logistic growth curve. See *logistic growth*.

logit analysis. A method of regression analysis with a limited dependent variable (usually binary) commonly used for discrete choice analysis. The logit model assumes a functional form for the regression in which $y_i^* = \beta x + u_i$, where y_i^* is unobservable, but can be deduced by assuming that an observable dummy, y, equals 1 if $y_i^* > 0$ and 0 otherwise. In logit analysis, u_i is a random variable with a logistical distribution. See *probit analysis*.

logit regression. See *logit analysis*.

log-likelihood. The logarithm of the value of the likelihood function used to obtain maximum likelihood estimators. See *likelihood function*.

FURTHER READING
Ramanathan (1992).

London Amendment. A 1990 amendment to the 1987 Montreal Protocol which speeded up the phased out reductions in the production of chlorofluorocarbons (CFCs) that destroy stratospheric ozone. See *Montreal Protocol*.

London Dumping Convention. First signed in 1972, and now ratified by 75 countries, the Convention on the Prevention of Marine Pollution by Dumping of Waste and Other Matter prohibits the dumping of specified wastes, including radioactive wastes.

longitudinal data. Data that includes repeated observations from the subject of study.

longline. A method of fishing by which lines with baited hooks are held at or below the surface by floats and weights. Longlining is a common method of harvesting large pelagic species, such as tuna and swordfish.

long-range transportation of air pollution (LRAP). The transportation of pollutants, such as sulfur dioxide, a great distance from the source of emissions.

long run. The period of time in which all factors of production can be varied.

long-run sustainable yield (LRSY). The culmination of the mean annual increment weighted by the area of all productive and utilizable forest land types in a forest stand. See *mean annual increment*.

lopping. The process of cutting tops of trees and branches after felling in order to increase the amount of slash left near the ground.

Lorenz curve. A graph which plots the relationship between cumulative income or wealth or some other variable (on the vertical axis) against the cumulative population (on the horizontal axis). If every person has the same income or wealth the curve is represented by a 45^0 line and the more unequal the distribution the greater the area between the Lorenz curve and the 45^0 line.

FURTHER READING
Ray (1998).

lotic. Term used for running fresh water.

Lotka–Volterra model. A predator–prey model governed by a two dimensional system of differential equations where the population of the two species fluctuates in a perpetual harmonic motion around an equilibrium that is never reached.

FURTHER READING
Grafton and Silve-Echenique (1997).

Love Canal. A community in upper New York state which was formerly a disposal site for chemicals and plastics. Concerns over seepage of toxic materials from the site into basements and soils, and the potential health hazards from such leakage, led to the area being declared a disaster area in 1980. The incident provided stimulus for CERCLA or the Superfund legislation. See *CERCLA*.

lowest observed adverse effects levels (LOAELs). A measure derived from animal and epidemiological studies. Exposure to a defined substance above its LOAEL may be associated with potential health problems.

low-intensity logging. Logging that removes less volume over a given area relative to more conventional logging techniques, such as clear-cutting.

Lucas critique. Named after Robert Lucas Jr., a Nobel laureate in economics, this states that the expectations, and thus behavior, of economic agents are likely to alter following changes in economic policies. Thus, economic policies should not be undertaken under the assumption that, on average, individuals have naïve expectations.

FURTHER READING
Romer (1996).

LULU.   See *locally unwanted land use.*

Lyme disease.   A bacterial disease transmitted by ticks to humans and named after the place it was first discovered on Long Island, New Jersey, in 1982. Increased populations of deer in North America have increased the number of ticks that feed on them and in consequence, the incidence of the disease in humans.

M

MAB. See *Man and the Biosphere.*

macroeconomics. The study of business cycles, economic growth and broad or large-scale economic problems using measures of economic aggregates and gross economic activity. For example, the study of unemployment is a branch of macroeconomics. See *microeconomics.*

"mad cow" disease. See *bovine spongiform encephalopathy (BSE).*

magma. Molten material found beneath the earth's surface.

Magnuson–Stevenson Act. A US federal act ratified in 1976 but changed in 1990, and renamed in 1996, that provides guidelines to fishery management councils. See *fishery management plans.*

malaria. A disease common in most tropical areas that is caused by several different parasites and is spread by the bite of certain types of mosquitoes. Malaria kills and debilitates millions of people, and control measures are often limited to the spraying of insecticides to reduce mosquito populations.

malnutrition. A generic term for a diet that lacks specific and necessary nutrients, such as protein. Millions of people in developing countries suffer from malnutrition, and it is a contributing factor to morbidity and mortality.

Malthusian. Term used to describe ideas or models that make pessimistic predictions about future resource availability. Named after the Rev. Thomas Malthus, whose eighteenth-century essay on population predicted that population growth would shortly be checked by food limits and resource shortages. See *cornucopian.*

mammal. Class of animals (including humans) that have several distinguishing characteristics, including mammary glands used for feeding young.

Man and the Biosphere (MAB). An interdisciplinary research and training program directed by the United Nations Educational, Scientific and Cultural Organization (UNESCO) to develop biosphere reserves and reconcile the sustainable use of natural resources with ecological integrity.

manganese nodules.   Deposits of iron and manganese found on the sea floor.

mangroves.   Trees found in estuarine environments in sub-tropical and tropical regions. Mangroves support important ecosystems and provide a "nursery" for some marine species, such as shrimp. In recent years, coastal developments (particularly in Asia) have led to concerns about the destruction of mangroves and the effects of this on the environment.

mantle.   The layer of rocks within the earth that lies above the core and beneath the earth's crust.

mare liberum.   Latin term for those parts of the sea or ocean that are freely navigable.

marginal abatement cost (MAC).   The change in the total cost of abatement of pollution or emissions following a marginal change in the level of pollution or emissions abatement. See *abatement*.

marginal cost.   See *marginal private cost*.

marginal cost of enforcement (MCE).   The additional cost of providing one more unit of enforcement effort.

marginal cost of exploration.   The cost of finding additional non-renewable resources. It is sometimes used as an indicator of resource scarcity.

marginal-cost pricing.   The pricing of goods such that the price per unit equals the marginal cost of producing that unit.

marginal external cost (MEC).   Change in costs imposed on the rest of a society following a marginal change in output.

marginal extraction cost.   Change in the total cost of extracting a resource following a marginal change in the amount extracted.

marginal physical product (MPP).   The extra output produced from a marginal change in the quantity of a variable input used in a production process.

marginal private cost.   Change in total private costs following a marginal change in output.

marginal product.   See *marginal physical product.*

marginal profit.   The difference between marginal revenue and marginal private cost.

marginal propensity to consume.   The proportion of a marginal increase in income that is consumed, rather than saved.

marginal propensity to save.   The proportion of a marginal increase in income that is saved, rather than consumed.

marginal rate of substitution (MRS).   The rate at which an individual is prepared to substitute one good for another while maintaining the same level of utility. It is the negative of the ratio of the marginal utilities of the two goods.

marginal rate of technical substitution (MRTS).   The rate at which one input can be substituted for another while maintaining the same level of output. It is the negative of the ratio of the marginal products of the two inputs.

marginal revenue.   The change in revenue following a marginal change in sales.

marginal social benefit.   The change in private and external benefits following a change in output.

marginal social cost.   Change in total private costs and the costs imposed on others (due to external costs) following a marginal change in output.

marginal stock effect.   The right-hand-side term in the fundamental equation of renewable resources defined by the ratio of the following partial derivatives

$$\frac{\partial \pi / \partial x}{\partial \pi / \partial h}$$

where x is the level or size of the renewable resource, h is the harvest, π is the current rent from the renewable resource. The marginal stock effect represents the marginal value of the biomass or resource stock relative to the marginal value of the harvest. See *fundamental equation of renewable resources.*

marginal utility.   The change in total utility following a marginal change in the consumption of a good or service.

marginal value.   The change in value attached to a marginal change in the quantity of an asset or resource.

marginal willingness to pay.   The willingness to pay, by a consumer, to acquire one more unit of a good.   The first derivative of *total willingness to pay* with respect to quantity.

marine fish farming.   The raising of fish in enclosed pens in ocean waters for commercial purposes.

market approach.   An approach in which free trade in the market determines the allocation of environmental goods or natural resources. The market approach is thought to achieve either a least cost or economically efficient allocation provided that property rights are well-defined and prices include all externalities.

market-based instruments.   Approaches to pollution control, or the management of natural resources, where the method of control is a quantitative right that can be traded. Tradable emission permits and individual transferable quotas are both market-based instruments.

market-clearing price.   The price in a market that ensures that the quantity demanded of a good or service exactly equals the quantity supplied.

market economy.   An economy where most of the decisions about what to produce and consume are decided by markets rather than planners. See *command economy*.

market environmentalists.   Term used to describe persons who believe in and promote the use of incentives and market-based instruments to address environmental problems.

market equilibrium.   A situation where the quantity supplied to a market exactly equals the quantity demanded such that there is no tendency for the price to change.

market failure.   The failure of a market to achieve the economically optimal allocation of resources caused by one or more of the following: increasing returns to scale in production that are not exhausted before equilibrium output is reached (e.g., monopolistic production), the presence of technological externalities in production or consumption, excessive uncertainty within markets, and a lack of a sufficient number of producers and consumers (i.e., thin markets), or incomplete property rights.

market price. The price at which a unit of a good or service can be purchased in the current market. The market price need not necessarily be a competitive price.

market risk. The inherent risk in holding equities. It cannot be eliminated by diversification of the portfolio of holdings.

market structure. The principal characteristics of a market including the barriers to entry, the number of buyers and sellers, the differences among products sold in the market, and other factors.

market stumpage prices. Prices for timber derived from arm's length transactions between the buyer and seller. Such prices, theoretically, should equal the market price of the wood embodied in the tree less all conversion costs.

Markov transition matrix. A matrix that defines the probability of going from one state of the world to another, and that can be used to model the dynamics of populations.

MARPOL. See *London Dumping Convention*.

Marrakesh Agreement, The. The conclusion of talks under GATT that started in Uruguay in 1986, and ended in Marrakesh in 1994. The agreement eliminated a wide range of tariffs in a number of different industrial sectors and established a framework for liberalized international trade. It also created the World Trade Organization (WTO). The agreement incorporated the need to address environmental concerns in the context of international trade. See *Uruguay Round, The*.

Marshallian demand. Demand for a good or service that is a function of output prices and income.

Marshallian surplus. The area under the Marshallian, or un-compensated, demand function between the price of a consumption good and the choke price of that good (i.e., the price that drives consumption to zero). Marshallian surplus can be considered a money metric measure of utility only when the marginal utility of income is unity for all goods. See *consumer surplus*.

marsupials. Mammals whose young are born live and then develop in a pouch or marsupium, such as a kangaroo.

mass spectrometer. An instrument used to measure the relative proportions or masses of different elements and their isotopes.

materials balance principle. Derived from the first law of thermodynamics, the materials or mass balance principle states that the matter/energy content of a material cannot be destroyed. Therefore, matter extracted from the earth (say ore) and processed must also be accompanied by waste. The more matter extracted and processed, the more waste is created.

mathematical expectation. See *expected value*.

matrix. A systematic array of numbers arranged in rows and columns.

maximin and minimax strategies. The same strategy, used in models in game theory, described from two different perspectives. Under the *maximin* strategy, all other participants in a game are trying to make a particular participant's payoff as low as possible. A particular participant then chooses the *maximin* strategy, which is the strategy that will yield the highest payoff for him/her, given the others' strategies and states of nature. The strategies pursued by the other participants, which are chosen to minimize the payoff to that particular party, are then collectively described as a *minimax* strategy.

maximum allowable concentration. The maximum concentration of a pollutant allowed in the workplace.

maximum likelihood. A widely used statistical method of estimation that maximizes the likelihood function from an observed sample for an assumed distribution of the residuals. The maximum likelihood estimators have the greatest probability of drawing the sample data used to derive the estimators. See *likelihood function*.

FURTHER READING
Ramanathan (1992).

maximum permissible dose. The maximum amount of ionizing radiation an individual should be able to absorb in a given period of time without negative health results.

maximum principle. Necessary conditions in optimal control theory that must be satisfied to ensure an optimal time path for the control and state variables in a dynamic optimization problem in continuous time. See *optimal control theory*.

FURTHER READING
Chiang (1992).

maximum residue limits. Limits set by food health and safety authorities that determine the maximum amount of a particular substance (such as a pesticide) allowed on a food (such as grapes).

maximum sustainable yield (MSY). The supposed maximum yield or harvest that can be taken or withdrawn from a renewable resource indefinitely without affecting the current level or stock of the resource.

maximum value function. See *optimal value function*.

Mayflower problem. The notion that the actual exploitation of non-renewable resources is governed by imperfect information and uncertainty such that lower quality resources may be used before higher quality resources. This notion is in contradiction to the theory of David Ricardo, the nineteenth-century British economist. Thus, the "pilgrim fathers" that came to the USA from England on the ship the *Mayflower* settled on poorer soils first as they did not know (nor could they use) better soils that existed to the west.

McKelvey box. A two-dimensional representation of the likelihood of the existence of mineral reserves (proven, probable, possible and undiscovered) versus the economic value of exploiting the reserves.

MEA. See *Multilateral Environmental Agreements*.

mean. See *arithmetic mean*.

mean annual increment (MAI). Average annual growth in timber volume in a given stand of trees expressed as an increase in volume per unit area.

mean squared error.   Expected value of the square of the difference between the true value of a parameter and its estimated value. It is equal to the variance plus the square of the bias.

measurement error.   Errors in the recording of data used in statistical analysis.

median.   The middle value from a sample or population where the observations are arranged from the lowest value to the highest.

mega.   Popularly taken to mean "very large",   the term is more precisely used as the prefix signifying a million or 1×10^6. (For example, a megabyte is one million bytes.)

megafauna.   Large animals (such as lions or gazelles) found in an ecosystem. Conservation efforts are often directed to megafauna rather than smaller-sized animals. See *fauna* and *charismatic fauna.*

megalopolis.   See *conurbation.*

melanism.   Term used for the appearance of dark skin, pigment or hair due to an abnormal increase in the production of melanin.

melanoma.   A malignant form of skin cancer that is linked to exposure to ultraviolet radiation. A decline in stratospheric ozone, in the absence of avoidance measures from sun exposure, will increase the incidence of melanoma.

meltdown.   The accidental melting of nuclear fuel in a reactor due to the breakdown of the cooling system.

membrane.   A pliable and thin tissue consisting of lipids and proteins that separates cells and other structures within plants and animals.

Mendel's laws.   Laws or results named after the plant breeder Gregor Johann Mendel which state that traits are inherited separately from each other and that recessive traits will only appear if there exist two recessive genes.

mercantilism.   Name given to a school of thought that had its origins in the sixteenth and seventeenth centuries in England. Mercantilists supported the notion of the state intervening to promote trade, such as the granting of trade monopolies.

mercury. A toxic metal found in many different chemical compounds and sometimes used as an amalgam in dental fillings. See *methyl mercury.*

mercury poisoning. See *Minamata disease.*

mesosphere. A part of the earth's atmosphere located around 50–80 kilometers above the earth's surface.

Mesozoic. Geological era that spans the period 245 million to 66.4 million years ago.

meta-analysis. A statistical method of analysis that combines the results of two or more independent studies.

metabolism. The combined physical and chemical processes within organisms that sustain life.

metagenesis. Name given for alternatively sexual and asexual reproduction.

metals. Any of the metallic elements and compounds characterized by a crystalline structure when solid.

meta-modeling. The creation of a general model from more detailed models that may differ in spatial and temporal scales.

metamorphosis. A fundamental change in the form of an organism during its life, such as in the life cycle of a moth from a caterpillar to its adult form.

meta-population. The population of a species as defined over its entire spatial distribution. A widely dispersed meta-population reduces the probability of extinction, as environmental shocks to one region or area that may lead to the extirpation of a species still allows the species to continue in another locale.

meteor. A rocky object that becomes visible upon entry to the earth's atmosphere.

meteorite. A meteor that reaches the earth's surface.

meteorology. The study of weather and the atmosphere.

meter. Unit of length equal to 3.28 feet.

methane (CH$_4$). One of the most important greenhouse gases with a global warming potential of 21. The major sources of methane emissions are coal mines and natural gas fields, livestock production, and decaying matter in landfills and bogs. See *global warming potential.*

methyl bromide. A widely used pesticide that has a high ozone-depletion potential.

methyl mercury. A toxic form of mercury created by the breakdown of mercury in the environment. Several important cases of mercury poisoning have occurred in people who had consumed foods (fish and shellfish) with high levels of methyl mercury. See *Minamata disease.*

methyl tert butyl ether (MTBE). A gasoline additive used in the United States by refineries to assist them in complying with state and federal regulations for "cleaner" gasoline. MTBE is an oxygenate that enables gasoline to burn more cleanly but it is highly water soluble, such that gasoline leaks and spills can contaminate water sources. Concerns over water pollution have led the State of California to phase out and commit to discontinuing its use by January 1, 2003.

metric system. Also called the SI, it is the set of standard international units of measure. Further details are available in Appendix 3.

microbe. Any organism that can only be seen with a microscope.

microclimate. A variation in the climate limited to a small spatial area, and often due to the physical landscape.

microeconomics. The study of economic behavior and problems at a household, individual or firm level. For example, evaluating the benefits of pollution permits to control sufur dioxide emissions is a part of microeconomics.

micrometer. A unit of measure equal to 1/1000 of a millimeter.

micronutrient. A nutrient required in very small quantities.

migratory species. A species that moves over considerable distances in its life cycle. Many fish and bird species are migratory. See *sedentary species.*

Milankovitch cycles.   A hypothesis developed by the mathematician Milutin Milankovitch. The Milankovitch cycles are regular changes or cycles in the earth's orbit, angle to the sun and precession of the equinoxes that affect the amount of solar radiation reaching the earth. The cycles have been used to explain the existence and timing of ice ages. See *ice age.*

mile.   Unit of length equal to 1.61 kilometers.

Minamata disease.   A disease named after a fishing village in Japan where its people were first diagnosed with a debilitating (and sometimes fatal) nervous disorder, which was subsequently found to be caused by the consumption of fish containing high levels of methyl mercury.

mineral.   Any non-living substance that is extracted from the surface or sub-surface of the earth by mining or quarrying.

mineralogical threshold.   A hypothesized threshold at which the cost of extracting certain metallic elements increases sharply due to a change in how the metals are chemically bound into various ores and rock bodies.

mineral reserve.   The amount of a mineral that is known to exist with a reasonable degree of certainty, and which can also be profitably extracted.

minimum environmental stock.   The minimum level of environmental goods and services required to support sustained economic activity.

FURTHER READING
Barbier (1993).

minimum-tillage farming.   An agricultural technique that minimizes the disturbance of the soil surface to reduce erosion.

minimum utilization rate.   A standard sometimes imposed on forest companies when harvesting timber where they have cutting rights but do not own the land. It specifies the maximum height of the tree stumps, and other rules, to encourage the most efficient utilization of the trees.

minimum utilization standard.   See *minimum utilization rate.*

minimum value function.   See *optimal value function.*

minimum viable population.   The   population   size   (or   stock   in fisheries)   below   which   death   and   emigration   exceed   birth   and immigration.   Populations that fall below the minimum viable population size will become locally extinct.

Miocene.   Geological epoch of time from 23.7 to 5.3 million years ago.

mist belt.   A   contiguous   area   with   persistent   precipitation   that condenses directly on vegetation.   See also *occult precipitation*.

mitigation.   The process by which the production of greenhouse gas emissions   is   constrained   by   anthropogenic   measures   such   as   by   fuel switching,   changes in land use,   use of more fuel-efficient technologies and carbon sequestration.

mixed economy.   An economy where the decisions about what goods and   services   are   produced,   how   many,   and   at   what   prices   are   made through   markets   as   well   as   interventions   by   government   and/or   state enterprises.

mixed forest.   A   forest   made   up   of   both   deciduous   and   coniferous species.

mixed-use development.   A   plan   that   involves   different   uses   of   the same site rather than dedication to a single purpose (e.g., forest lands that support wildlife habitat and recreation, not just timber production).

mobile source.   A   source   of   pollution   that   can   be   non-stationary,   such as an automobile. See *stationary source*.

mode.   Most frequent observation in a sample or population.

modeling.   A   method   of   inquiry   by   which   a   simplified   abstraction   is used   to   represent   and   predict   phenomena.   Economic   models   often consist of equations, but a model may also be written in sentences or represented by graphs, flow charts or other means.

modus operandi.   Latin term meaning "the way things are done".

modus vivendi.   Latin term for a working compromise between parties.

moisture seasonality.   Seasonal patterns in the availability of moisture over the year.

molecule. At least two atoms of one or more elements chemically combined. For example, carbon dioxide is a molecule that consists of one carbon atom and two oxygen atoms.

mollusks. A common and widespread group of marine and freshwater animals that are also found on land. Mollusks are capable of producing a hard shell from secretions of calcium carbonate, and have a tongue-like organ. Mollusks include scallops and mussels, as well as snails.

monetary policy. Policy of the central bank regarding the control of the money supply and its effects on output, inflation and the exchange rate.

monoculture. Single-species cropping.

monopolistic competition. A market structure where there are many buyers and sellers, ease of entry and exit, and in which each firm produces a slightly different product.

monopoly. A market structure where there is only one seller that can influence the price of the good it sells by changing the quantity it chooses to produce.

monopsony. A market structure where there is only one buyer in the market that can influence the price it pays for a good or service by changing the quantity it chooses to purchase.

monotypic genus. A genus of plants or animals containing only one extant species. Humans are a monotypic genus.

monsoon. Commonly refers to the reversal of wind directions that occurs at different times of the year and can extend from east Africa to Asia. In India, the summer monsoon winds blow from the south-westerly direction bringing increased precipitation, which is critical to ensuring farming success.

montane. A sub-alpine environment characterized by a cold winter and coniferous trees.

Monte Carlo methods. Data-simulation approach to testing models and hypotheses.

montmorillonite.   A clay derived from the weathering of rocks that is noted for swelling in wet periods and contracting in dry periods.

Montreal Process.   An initiative started by the Government of Canada, which hosted a meeting in Montreal in September 1993 to develop a series of criteria and indicators for sustainable forest management. The final meetings were held in Santiago, Chile, in 1995 and produced two documents: the Santiago Declaration and "Seven Criteria and Associated Quantitative and Qualitative Indicators". See *Helsinki Process*.

Montreal Protocol on Substances that Deplete the Ozone Layer.   An agreement first signed in 1987 in Montreal, Canada, and often called the Montreal Protocol, under which signatories agreed to reduce their production of ozone-damaging chlorofluorocarbons (CFCs) by half by the year 2000, and which continued the process begun in 1985 with the Vienna Convention for the Protection of the Ozone Layer. The Protocol was subsequently strengthened in 1990, and again in 1992, with the London and Copenhagen Amendments, under which all CFC production would stop by 2000.

moral hazard.   A problem of asymmetric information whereby the actions of one party to a transaction are unobservable. For example, if people were able fully to insure their valuables (with no deductibles) they might be less inclined to undertake actions that would reduce the probability of theft than if they were obliged to cover some of the costs of loss. See *asymmetric information*.

morbidity.   The incidence of disease. A morbidity rate is usually defined as the number of reported cases of a disease in a population per 100,000 people in a given period of time.

more-is-better property.   The assumption that larger quantities of any good generate higher levels of utility from the consumption of that good. Also known as local non-satiation.

mortality rate.   The death rate of a population over a given period of time. Human mortality rates are defined as the number of deaths per 100,000 people per year.

most-favored nation status.   A status accorded to countries by the US Congress in order to promote trade between the USA and that country. Pressure groups have lobbied to remove some countries from the most-favored list so as to pressure certain states to accept democracy, human rights and environmental causes.

Mount Pinatubo. A volcano in the Philippines that erupted in 1991 and which pushed a large amount of debris into the stratosphere. This material reduced the solar radiation reaching the earth's surface and led to a cooling of the northern hemisphere in 1991 and 1992.

Mount Tambora. Site of a major volcanic eruption in 1815, in what is today known as Indonesia, that had a major short-term impact on the earth's weather and resulted in significant cooling.

moving average (MA) process. A stochastic process by which the current value of a variable is a function of past errors. An MA process of order n is defined by

$$x_t = \alpha + e_t + a_1 e_{t-1} + a_2 e_{t-2} + \ldots + a_n e_{t-n}$$

where x_t is the variable in time t and e_t is an unobservable and uncorrelated random error with a zero mean and a constant variance.

MTBE. See *methyl tert butyl ether*.

mud flats. Low-lying areas along the shoreline that are exposed during periods of low tide and typically found in estuaries.

Mullerian mimicry. Biological mimicry whereby poisonous or unpalatable species mimic each other in appearance, thus reinforcing the trait and accentuating their ability to avoid predation. See *Batesian mimicry*.

multicollinearity. Correlation between independent variables in a multiple regression that make regression coefficients insignificant, although collectively the model may have a high goodness of fit.

multi-factor productivity (MFP). See *total factor productivity*.

Multilateral Environmental Agreements. Agreements worked out between different countries over environmental issues ranging from regional to international problems. See *Convention on International Trade Endangered Species* and the *Montreal Protocol*.

multilateralism. Term used to describe a co-ordinated and co-operative approach to international relations to promote trade and peace and to address global and regional environmental problems among more than two different countries.

multi-media pollution.   Pollution that occurs across several vectors that may include air, water and soils.

multinational corporations (MNCs).   Companies that have plants and operations in several different countries. Some multinationals are so large that their sales dwarf the gross domestic products of many smaller countries.

multiple chemical sensitivity (MCS).   Term used to describe a complaint where patients become ill from contact with very small amounts of chemicals that normally would not be a cause for concern.

multiple-purpose trees (MPTs).   Trees, especially those used in agroforestry endeavors, that provide more than one important resource including fodder, fuelwood, timber, windbreaks, and, in the case of leguminous MPTs, nitrogen for the soil.

multiple regression.   Analysis where a single dependent variable is estimated from at least two independent or explanatory variables. See *regression analysis.*

multiple use.   Term used to describe the joint use of habitats or ecosystems. In forestry, multiple use may involve some tree harvesting combined with the maintenance of corridors to preserve ecological integrity.

Multiple-use management.   The management of a single resource (e.g., a forest) for multiple and jurisdictionally overlapping uses (e.g., camping, hunting and timbering).

multiplier.   An economic term that describes the change in income following a change in expenditure.

municipal solid waste (MSW).   Waste not normally considered as hazardous that is typically disposed of in city landfills.

municipal tipping fee.   A payment charged by a municipality for the right to dump refuse at a site.

mutagen.   Any agent capable of causing genetic mutation, including various chemicals and short-wave radiation.

mutation.   A significantly different, but inheritable, change in the genome of an organism.

mutatis mutandis.   Latin term meaning "with the necessary changes".

mutualism.   A mutually beneficial relationship between at least two different species. For example, mutualism exists between flowering plants and bees because the flowers provide nectar and the bees help pollinate the flowers.

myopic behavior.   Behavior, especially economic behavior, that does not take into account associated future costs and benefits of actions or event and focuses upon a short-term horizon.

N

NAFTA. See *North American Free Trade Agreement.*

Nash equilibrium. A concept in game theory that describes a game outcome where an individual participant has no reason to change his or her own strategy, given the strategy chosen by all the other participants.

natality rate. See *birth rate.*

national action plan (NAP). A country's stated objectives for pollution control, and the policies designed to meet those goals.

national ambient air quality standards (NAAQS). US federal standards that set the levels of legally permitted concentrations of defined pollutants in a defined volume of air.

national debt. The cumulative borrowing by a government. Debts in excess of 60 percent of the gross domestic product are considered to be a potential economic problem.

national emission standards for hazardous air pollutants (NESHAP). US standards for point sources of emissions of potentially hazardous pollutants.

National Environmental Policy Act (NEPA). Legislation passed in the USA in 1970 that created both the Environmental Protection Agency (EPA) and the requirement for federal agencies to evaluate the potential impact from agency actions or proposed legislation on the environment through an Environmental Impact Statement (EIS), followed by an Environmental Impact Assessment (EIA), if there is found to be an impact. The formal preparation and incorporation of environmental assessments into decision-making has subsequently been adopted by a number of countries.

FURTHER READING
Callan and Thomas (2000).

national forests. Forestlands held by the government for the benefit of the country as a whole. In the USA, national forest-lands refer to those lands managed by both the US Forest Service and the Bureau of Land Management, all of which are owned by the federal government.

national income.   The total income of all residents in an economy, less the costs of capital depreciation.

national income and product accounts (NIPA).   A set of accounts that measure economic activity in an economy, including estimates of national income, output and expenditures.

nationalization.   The appropriation of companies or industries by the state, with compensation to the previous owners, so as to create state-owned enterprises.

National Oceanographic and Atmospheric Administration (NOAA). US agency charged with the federal management of oceans and the marine environment.

national pollutant discharge elimination system (NPDES).   A US system designed to control the discharge of effluents from designated industrial polluters.

national primary drinking water regulations (NPDWR).   US regulations designed to ensure domestic drinking water meets designated minimum health standards.

national priorities list (NPL).   A US list of hazardous waste sites that have the highest priority, in terms of control, so as to prevent negative environmental effects.

native rights.   The human and property rights of native or aboriginal people to use their traditional land and preserve their traditional way of life.

native species.   A species that has been found in a particular region for a very long period of time.

natural attenuation.   Naturally occurring processes that can reduce the mass or toxicity of contaminants in polluted soil or water.

natural capital.   A collective term for all aspects of the environment, including natural resources.

natural gas.   Gas formed by hydrocarbons, which consists mainly of methane, ethane and propane.

natural increase. The positive difference between the birth rate and death rate in a population.

naturalism. 1. A reverence for natural processes and objects. 2. The belief that scientific inquiry can explain processes.

natural monopoly. Term used to describe industries that exhibit increasing returns to scale such that one producer or firm is able to produce at lower cost per unit than if the same quantity were produced by several different firms.

natural radioactivity. The background radiation level given off by the decay of naturally occurring isotopes.

natural rate of unemployment. See *non-accelerating inflation rate of unemployment.*

natural resource economics. A sub-discipline of economics concerned with the management and use of renewable and non-renewable resources.

natural resources. Resources in the environment that include both renewable (such as fisheries) and non-renewable resources (such as oil and gas deposits).

natural selection. Process described by Charles Darwin where traits that are best suited to the existing environment tend to favor the reproductive success of individuals with such traits. See *Darwinian evolution.*

FURTHER READING
Darwin (1859).

natural subsidy. The availability of natural energy resources in a form that is easily accessed and used by the human economy. Energy resources, such as East Texas petroleum, that require little labor and capital to be extracted and processed are said to have a natural subsidy relative to other more costly energy resources, such as Alaskan petroleum.

FURTHER READING
Daly (1996).

nature-based tourism.   Tourism that has, as its main attraction, access to nature. Nature tourism differs from ecotourism in that nature tourism does not necessarily meet criteria designed to ensure low environmental impacts.

Nature Conservancy.   A non-governmental organization that focuses its efforts on purchasing private land so as to conserve endangered ecosystems and areas of particular beauty or natural significance.

nautical mile.   Unit of distance equal to 1.151 miles, or 1.852 kilometers.

NDVI.   See *Normalized Difference Vegetation Index.*

necessary conditions.   Conditions that must be satisfied at the optimum, or the solution to an optimization problem.

negative externality. An action which imposes costs on others but the person(s) undertaking the action do not account for these "external" costs in their decision making. For example, a firm that pollutes a river without consideration of the effects on downstream users is causing a negative externality. See *externality* and *positive externality.*

negative feedback.   See *feedback.*

negligence.   Legal term used to describe the behavior of individuals, firms or governments where due diligence has not been undertaken regarding decisions or actions.

nektonic.   Term used to describe any sea creature that can move using its own power.

neoclassical economics.   The most popular mode of analysis of economics at present. It is based on the premises of rational behavior by economic agents, optimization by individuals and firms who do the best they can given their constraints, and equilibrium in the sense that systems will not move from a state of rest without a shock.

neoclassical growth theory.   Models used to explain economic growth using economic aggregates, paying particular attention to capital/output and capital/labor ratios. Neoclassical growth models imply that an increase in the savings rate, population growth or changes in the level of technology do not affect the steady-state per capita growth rate in output which always equals zero in the long run.

FURTHER READING
Barro and Sala-i-Martin (1995).

neoteny.   Term for juvenile or immature characteristics that are retained by an organism in its adult form.

NEPA.   See *National Environmental Policy Act.*

neritic.   Marine zone usually defined as less than 200 meters in depth.

net domestic product (NDP).   The gross domestic product less an allowance for depreciation of the capital stock.

net energy.   The work done from an energy source less the work needed to find, extract, process and use the energy source.

net national product (NNP).   The gross national product less an allowance for capital depreciation.

net national income (NNI).   The net national product plus subsidies but less indirect taxes (such as value-added taxes).   See *net national product.*

net present value (NPV).   The sum of the discounted net benfits (benefits minus costs) in each time period, over the lifetime of the investment. See *present value.*

net primary productivity (NPP).   The total energy per time period stored by photosynthesis in an environment less the total amount of energy used by the plants in respiration. NPP provides a measures of the biological productivity of an ecosystem.

net production.   The energy of an organism devoted to growth and reproduction.

net production efficiency.   Proportion of total energy consumed by an organism that is devoted to reproduction and growth.

net social surplus.   The net value to society from the production (extraction) and consumption of a good or service measured as the total benefits from consumption minus the total costs of production (extraction).

netting. A sub-program of emission credit trading set up in the USA to allow some flexibility in meeting air pollution control targets. It allows a firm to modify its plant without going through an involved new source review. Netting allows a firm to increase its emissions for a given pollutant from one point provided that there is a corresponding decrease in emissions at another point so that net emissions do not increase overall for the firm.

FURTHER READING
Tietenberg (1996).

neurotoxin. Any substance that affects the functioning of the nervous system.

neutral tax. A tax that, in theory, does not distort economic behavior. A lump-sum tax that is paid irrespective of the actions of individuals or firms is considered to be the closest to a neutral tax.

neutron. An atomic particle that contains no electric charge. This is the part of the atom that is dislodged in nuclear fission. See *nuclear fission* and *nuclear fusion*.

new source performance standards. USA based technological standard of pollution control to which new plants must conform.

newton. A measure of force equal to a mass of one kilogram that has an acceleration of one meter per second per second.

NGOs. See *non-governmental organizations*.

niche. The function or part played by an organism within an ecosystem.

niche requirements. The biological needs of a particular species within the environment such as the availability of food, habitat, as well as the constraints placed upon it by physical characteristics such as temperature, water salinity, and availability of other resources.

NIMBY. Acronym for Not in My Back Yard. The term is used to describe individual and community efforts to prevent waste disposal plants and other undesirable facilities near their own communities or places of employment. See *NIMTOO* and *locally unwanted land use.*

NIMTOO. An acronym for Not In My Term Of Office, used to describe public officials' unwillingness to support controversial developments because of localized public concern in the area where the development will take place. See *NIMBY* and *locally unwanted land use.*

nitrates. A natural salt formed by micro-organisms in the soil that can be used by plants as a source of nitrogen. Excessive use of nitrogenous fertilizers has contaminated groundwater sources because nitrates can leach from the soil into aquifers.

nitrites. A salt formed by micro-organisms in the conversion of nitrogen into nitrates.

nitrogen. The principal constituent of the earth's atmosphere, it is also an essential element required by plants and animals.

nitrogen cycle. The biochemical processes through which atmospheric nitrogen is fixed and assimilated in the biosphere via soils and lakes, then returned to the atmosphere through denitrification.

nitrogen dioxide (NO_2). A very common atmospheric pollutant that is formed by automobiles when nitric oxide combines with oxygen in the combustion process. Nitrogen dioxide can produce tropospheric (surface) ozone in the presence of sunlight.

nitrogen-fixing. The act of removing nitrogen from the earth's atmosphere or from mineral compounds in the soil or fresh water by bacteria and algae and converting them into compounds available for use by other biological organisms.

nitrogen-fixing bacteria. Bacteria found in leguminous plants that convert atmospheric nitrogen and nitrogen found in soil and freshwater into compounds that can be used by plants and other organisms.

nitrogen oxides. See *oxides of nitrogen.*

nitrous oxide (N_2O). A gas used as an aerosol propellant and which is generated as a byproduct in the manufacture of fertilizers and the production of fossil fuels. Nitrous oxide is an important greenhouse gas with a global warming potential of 310.

NOAA. See *National Oceanographic and Atmospheric Administration.*

NOAA panel. The distinguished panel of experts convened by the National Oceanographic and Atmospheric Administration (NOAA) in 1992 to examine the contingent valuation methods (CVM) of non-market valuation. Its findings provide a set of guidelines on the approaches that should be followed in CVM studies.

Noah's ark problem. The problem that not all endangered species, or species at risk, may be saved. The problem poses difficult choices as it may be necessary to decide which species should be given priority in terms of preservation.

noise and number index. An index for measuring noise pollution at or near airports, based on noise levels and the number of aircraft landing and taking off at a given point in time.

nominal value. Value, in terms of currency, before considering the rate of inflation. See *real value.*

non-accelerating inflation rate of unemployment (NAIRU). The supposed unemployment rate below which a reduction in unemployment would result in an acceleration in the inflation rate. See *consumer price index.*

FURTHER READING
Romer (1996).

non-Annex I countries. Mainly developing countries that are not included in Annex I of the Framework Convention on Climate Change, and which are not obligated to control their levels of GHG emissions to 1990 levels by the year 2000. See *Framework Convention on Climate Change.*

non-Annex B countries. Mainly developing countries that are not included in Annex B of the Kyoto Protocol, and thus are not obligated to control their level of emissions by the target period 2008–12. See *Kyoto Protocol.*

non-attainment area. A term in the USA for an air quality control region that fails to meet the designated air quality standards.

non-biodegradable materials. Materials that will not naturally decompose in the environment within a reasonable time-frame.

non-compliance. Term that indicates a firm or individual is not currently meeting a defined environmental regulation or standard.

non-consumptive benefits. Benefits from resources or the environment that do not arise from any direct extraction or harvesting. See *non-consumptive use.*

non-consumptive use. Use of the environment or natural resources, such as photography, that does not detract from the benefits these provide.

non-cooperative game theory. The branch of game theory that involves models of behavior where individual participants are assumed to choose actions that will only maximize their own well-being.

non-diversionary water uses. Those uses, such as fishing, boating, or the provision of habitat, that require in-stream water flows.

non-exclusive. An asset or resource which others cannot be prevented from using. See *exclusivity.*

non-extractive use value. The consumptive value of a good *in situ,* in nature.

non-fuel minerals. Mineral resources excluding those used for the production of energy (e.g., petroleum, coal and uranium) but including aluminum, copper, and other metals and ores.

non-governmental organizations (NGOs). Registered non-profit organizations that may be funded by individual members, industry and/or a country, but which are not directly controlled by the government. NGOs play an important role in providing linkages and technical assistance to developing countries and are influential in providing information and influencing policies on the environment in many countries.

non-jointness in inputs. A production relationship where different outputs can be produced using separate production functions.

non-linear. Any relation or function where the dependent variable does not vary proportionally with changes in other variables of the model.

non-market damages. Costs imposed on the environment but which are not reflected in the price of goods and services. For example, the price of air-conditioning units which used chlorofluorocarbons (CFCs) as a refrigerant did not include the environmental costs associated with stratospheric ozone depletion.

non-market valuation.   The attempt to place a monetary value on goods, especially public or environmental goods, that are not traded directly in the market. See *contingent valuation* and *hedonic analysis*.

non-material values.   Those human values, such as spiritual beliefs, which are not associated with material goods or money.

non-pecuniary rewards.   Benefits that are not financial in nature, such as goods or improved social status.

non-point source.   Source of pollution where emissions and discharges cannot be traced to a specific point or location. Agricultural run-off is a common source of non-point pollution. See *point source*.

non-price rationing.   Restricting the quantities available for consumption by means other than price, such as rationing.

non-renewable resource.   A natural resource, such as fossil fuels, that has a finite stock and cannot be renewed.

non-rivalness.   See *rivalrous*.

non-stationary.   See s*tationary state*.

non-tariff barriers.   Measures or procedures, such as quotas, which are placed by countries on the importation of foreign goods, are not tariff-based, and that effectively restrict imports. Examples might include product safety or labeling requirements designed to either exclude foreign products or impose higher costs on them due to safety or environmental concerns.

non-timber forest products.   Natural products, other than timber, that can be harvested from a forest stand, usually without significant damage to standing trees. Examples include latex, syrup, vines, fibers, fruits and game.

non-timber resource values.   Values that emanate from forests but do not require the harvesting of forest stands. Non-timber resource values include values that come from the existence or enjoyment of biological diversity, recreation, fisheries, wildlife, water resources and water quality, minerals, and cultural and historical resources.

non-timber values. Value of the forest separate from the commercial or timber value of the trees. For example, forests provide amenity values for recreational purposes that are non-timber values.

non-use benefits. Benefits from resources or the environment that do not involve its direct use (consumptive and non-consumptive) such as the value placed on the existence of blue whales. See *non-use value.*

non-use value. Also intrinsic value, existence value, or preservation value. The total willingness to pay to preserve, maintain, or possess a natural resource or environmental good in excess of the willingness to pay to use that good *in situ.* Non-use value = total economic value - use value.

non-zero sum game. A game or a set of transactions whereby all players or persons participating have the potential to be better off. See *zero sum game.*

no observed adverse effects levels (NOAELs). Measures obtained from animal studies or epidemiological studies that are used as guidelines for determining maximum levels of exposure and risks associated with different exposure levels.

No-regrets policy. The notion that mitigation of pollution generates benefits beyond those associated with the immediate problems of pollution. For example, lowering the emissions of chlorofluorocarbons (CFCs) reduces greenhouse gas emissions that may contribute to global warming, but it also reduces the anthropogenic deterioration of stratospheric ozone.

normal distribution. A distribution where the mean, mode and median are identical, so that is is symmetrical around the mean. If the frequencies of a normally distributed variable are drawn on the vertical axis and the values are given on the horizontal axis, the resulting curve or distribution would resemble the shape of a bell where the top of the bell is the value of the mean, mode and median. Many characteristics of a sufficiently large enough population (such as height of individuals) can be characterized by a normal distribution.

normal good.   A good where the demand varies directly with income. See *inferior good*.

Normalized Difference Vegetation Index (NDVI).   An index of ground cover used to characterize vegetative canopy, ground cover and biomass. NDVI has been used extensively to describe remotely sensed data, especially from AVHRR sources. See *AVHRR*.

normal profit.   The difference between total revenue and total cost, including all opportunity costs. See *opportunity cost* and *profit*.

normative analysis.   See *normative economics*.

normative economics.   A term used to describe economic analysis which focuses upon "what should be", such as recommendations for economic policy, and that involves some value judgements.

norms.   Commonly accepted forms of behavior within a society. Changing norms from "consumption and disposal" to "consumption, recycling, and reuse", can improve overall environmental quality.

North American Free Trade Agreement (NAFTA). An   agreement signed between the USA, Canada and Mexico in 1992 to promote free trade among the three countries. The agreement includes various environmental clauses and commits the signatories to promoting sustainable development.

FURTHER READING
Hufbauer and Schott (1998).

"No Surprises" clause.   Under the Endangered Species Act in the USA, a landowner who is correctly implementing a Habitat Conservation Plan (HCP) is guaranteed to be excluded from any additional burden imposed by changes in regulations or restrictions on the use of their property. See *Habitat Conservation Plan*.

nuclear fission.   The splitting of uranium atoms by neutrons to produce lighter elements with at least two neutrons. Fission is the basis of the nuclear power industry.

nuclear fusion.   The fusing of two lighter elements (deuterium and tritium) to form hydrogen and a free neutron.

nuclear waste.   Radioactive material generated from nuclear tests and the operation of nuclear power stations. Low-level nuclear wastes can be disposed of relatively easily but highly radioactive material requires very costly methods of storage that, because of the long half-life of some waste, must last for thousands of years. See *radioactive waste.*

nuclear winter.   A hypothesis that a major nuclear conflict would spread so much particulate matter in the atmosphere that surface temperatures in higher latitudes would fall such that summer would be as cold as a normal winter.

FURTHER READING
Dotto (1986).

null hypothesis.   The hypothesis to be tested in a statistical test. For example, if we are testing whether the means of two populations are equal, then the null hypothesis, denoted by H_0, is that the means are equal and the alternative hypothesis, denoted by H_1, is that the means are not equal. The hypothesis test allows us to either fail to reject or reject the null hypothesis. If we reject the null hypothesis then we accept the alternative hypothesis. See *hypothesis* and *alternative hypothesis*.

nutrient concentrations.   The availability of organic and inorganic material from both external and internal sources that can be utilized by plants and animals within an ecosystem.

nutrient cycle.   A cycle in which minerals and nutrients are transported via soils, water, air and organisms.

O

objective function. A function that is maximized or minimized in a model.

Occam's Razor. A principle of scientific thought that holds simplicity and parsimony in explanation superior to complexity, all else equal.

occult precipitation. A mist-like form of precipitation in which condensation forms directly on vegetation.

oceanography. The science and study of oceans, including both biotic and abiotic systems.

oceans. The contiguous bodies of salt water that encompass about 70 percent of the earth's surface and provide a home to an abundance of marine life. Oceans also moderate temperatures along coastal areas.

OECD. See *Organization for Economic Cooperation and Economic Development.*

offsets. A sub-program of emission credit trading set up in the USA to allow some flexibility in the meeting of air pollution control targets. Offsets apply only to non-attainment areas allowing for new point sources of pollution provided that they are more than compensated for by reductions in emissions from existing sources.

FURTHER READING
Tietenberg (1996).

off-shore oil drilling. The extraction of petroleum and petroleum-related resources from the ocean floor by drilling.

off-site waste management. The treatment of waste at a location different from that where it was generated.

offtake. Harvest.

Ogallala. An alluvial fan system in the USA that stretches from South Dakota to Texas in the south and includes parts of Wyoming, Colorado, Nebraska, Kansas and Oklahoma. The system is a very large and widely used aquifer where the withdrawal rate greatly exceeds the rate of regeneration.

oil shale. Petroleum deposits in which the oil is bound into the sedimentary rock and has to be released through mechanical means.

oil spills. The accidental emission of petroleum or petroleum-related products into the environment.

old growth. Term used to describe forest of a climax community that may contain live trees of various ages from seedlings to very old and dead trees, but which is dominated by mature trees. Depending upon the environment, an old-growth forest may take hundreds of years to develop and often has high non-timber and timber values.

Oligocene. Geological epoch 36.6 to 23.7 million years ago.

oligopoly. A market structure with a few sellers, each of whom can influence the market price.

oligotrophic. A body of fresh water that is low in nutrients and productive capacity.

OLS. See *ordinary least squares*.

omnivorous. Animal that can feed on both plants and other animals, such as a bear.

one-shot game. Models used in game theory where the participants do not interact with one another again after they have completed the game. See *repeated games*.

one-tailed test. A statistical test that only uses one critical region when testing a null hypothesis. For example, $H_o : b_1 = 0$ is a one-tailed test if the alternative hypothesis is $b_1 < 0$. See *two-tailed test*.

open access. Term used to describe the lack of property rights over resources whereby no controls exist over the number of users or how much of the resource they may extract. Open access almost always leads to biological and economic overexploitation of natural resources.

open-access externalities. The imposition of costs upon others by individual actions, created by the open-access nature of the resource. See *externality* and *prisoner's dilemma*.

open-access resources. See *open access*.

open-cast mining. Also known as strip mining, the removal of overlying surface material to access an underground mineral deposit.

open economy. An economy where no restrictions exist in terms of imports and exports, investment and currency transfers. See *closed economy*.

open forest. Forest that has tree crown cover of between 10 and 30 percent. This would also include savannas.

open loop. A control system whereby the control variable does not respond to changes in the variables of the system.

open-loop recycling. The reuse of residual material products in other productive activities. Examples include the use of ground-up glass in road surface materials and shredded rubber in flooring products. See *closed-loop recycling*.

open system. A term used to define a physical system where there is an exchange of energy and mass or matter with the environment. Most natural environments are open systems. See *closed system*.

operable forest. A forest, or portion of a forest, in which timber can be profitably harvested under current market conditions.

opportunity cost. The cost of an economic activity in terms of the other goods and services, in the next best alternative, that must be forgone to be able to undertake the activity. For example, an opportunity cost associated with the installation of a pollution control device may be the forgone production that would otherwise would have occurred if the firm had, instead, invested in additional capital equipment.

optimal control theory. A set of mathematical methods and a theory developed to model dynamic systems that is closely related to the calculus of variations. The theory is used to solve problems where an objective function is maximized over time subject to set of constraints. A basic result of the theory is Pontryagin's maximum principle, which is concerned with the necessary conditions for optimality.

FURTHER READING
Chiang (1992) and Takayama (1985).

optimal rotation. The age at which a stand of replanted trees is harvested that yields the highest expected value of the forest. The optimal rotation depends on whether or not the land will be replanted after harvesting, and whether or not non-timber values are accounted for. See *Faustmann rotation* and *Hartman rotation.*

optimal value function. The function formed by substituting the optimized choice variables of an optimization problem into the original objective function.

FURTHER READING
Grafton and Sargent (1997).

optimal yield. A term that is not well-defined, but is frequently used to refer to a level of harvest from a renewable resource that will maximize overall benefits to society.

optimization. Term used for the maximization or minimization of an objective function subject to a set of constraints. See *objective function.*

optimum yield. See *maximum sustainable yield.*

option price. The maximum amount an individual would pay to preserve the option to use a resource in the future even if the individual is unsure whether she will demand that resource in the future. The option price is equal to the expected use value plus the option value.

option value. A value, beyond an expected use value, derived from the willingness to pay for the option of using a resource at a future date.

FURTHER READING
Wiesbrod (1964).

order. A classification of organisms by family. For example, the order of Primates includes the family of Hominidae which includes humans. See *family.*

order of a matrix. The number of rows and columns in a matrix. A matrix of order (2x3) has two rows and three columns.

order of integration. The number of times a non-stationary variable needs to be differenced in order to make the differenced variable stationary. The order of integration is written as I(n) where n represents the differences required to ensure stationarity. See *differencing* and *stationarity*.

ordinal utility. A preference structure in which individuals are assumed to be able to rank different combinations of goods and services. Ordinal utility forms the basis for demand theory in economics.

ordinary least squares (OLS). A statistical method of analysis for estimating parameters in a relationship between a dependent variable and one or more independent variables. OLS estimates the parameters by minimizing the sum of the squared deviations between the estimated relationship and the actual observations. See *regression analysis.*

Ordovician period. Geological period of time from 505 to 438 million years ago.

organic compounds. Material or substances containing carbon bound with hydrogen.

organic farming. A set of agricultural practices in which no artificial fertilizers, hormones, pesticides, herbicides are used. Organic farming is becoming increasing popular in developed countries as consumers are becoming more concerned over the potential health hazards of chemical residues in food.

organic fertilizer. A mixture derived from organic constituents made up of various compounds that aid in plant growth.

organic matter. Commonly defined as organic material found in soils that, in general, improves fertility and helps to retain moisture.

organic soil. Soil containing greater than 30 percent organic matter.

organism. A single animal or plant that is capable of functioning as an individual unit.

Organization for Cooperation and Economic Development (OECD). Formed in 1961, it is the so-called "Rich Man's Club" of countries. Its mandate is to foster economic growth, financial stability and high employment in member countries.

Organization of Petroleum Exporting Countries (OPEC). Formed in 1960, it consists of 12 major petroleum exporting countries: Algeria, Gabon, Indonesia, Iran, Iraq, Kuwait, Libya, Nigeria, Qatar, Saudi Arabia, the United Arab Emirates and Venezuela. The oil price shocks of the 1970s were, in large measure, due to the collusion of OPEC members.

organochlorides. A large number of organic chemical compounds containing chlorine including PVCs, PCBs, CFCs and even DDT. Organochlorides are very persistent in the environment and accumulate in fatty tissue, and some are known carcinogens.

organochlorines. See *chlorinated hydrocarbons*.

orogenesis. Tectonic movements that build mountains. Orogenesis has been used by some scientists to explain the timing and existence of ice ages.

orthogenesis. The notion that evolution's progression over time leads to more complex organisms.

orthogonal. A statistical term that indicates independence of two variables.

otolith. Bone in the ear of fish that is frequently used for determining the age of fish.

Our Common Future. The 1987 final report of the World Commission on Environment and Development that is commonly known as the Brundtland Commission after its Chair, Gro-Harlem Brundtland. The report was a major catalyst for governments to begin implementing policies to encourage sustainable development.

FURTHER READING
World Commission on Environment and Development (1987).

outfall. A point source of water pollution.

outlier. An observation in a sample that differs markedly from other observations.

output controls. Controls on the individual output of resource users. Output controls, in contrast to input controls, attempt to change the incentives faced by resource users by giving them a long-term interest in the resource. See *Individual Transferable Quotas* and *input controls*.

outputs. The end result of a production process whereby inputs, or factors of production, are combined to produce a good or asset. See *inputs*.

overcapacity. Capacity in excess of that required for a desirable level of output. Overcapacity is a common problem in many fisheries due to a lack of well-defined property rights over the resource. See *capacity*.

overexploitation. Biological or economic over use of a natural resource. A common reason for overexploitation is a lack of well-defined property rights.

FURTHER READING
Devlin and Grafton (1998).

overfishing. Fishing at a level that exceeds what is desirable from either a biological or economic perspective.

overhead. See *fixed costs*

overland flow. The flow of water outside of a water channel.

overlapping generations (OLG). A class of economic models in which the consumption of every generation, all equally long-lived, occurs simultaneously with at least one other generation.

overparameterized. A statistical model where the number of parameters to be estimated exceeds the number of observations.

overstorey. The layer of vegetation in a forest formed by the canopy of the dominant trees.

overvaluation. Placing too high an economic value upon a certain activity or resource without taking into account all of the associated economic costs.

oxides of nitrogen (NO$_x$).   A collective term for the pollutants nitric oxide, nitrous oxide and nitrogen dioxide. NO$_x$ is partly responsible for photochemical smog and, in the stratosphere, can contribute to depletion of ozone.

oxygen.   A gaseous element that, after nitrogen, is the largest component of the earth's atmosphere.   Oxygen is a component of all organic compounds and is present in most organic processes.

oxygen cycle.   The biological cycling of oxygen through the atmosphere and the biosphere.   The primary mechanisms of the cycle take place through the uptake and release of oxygen through green plants.

ozonation.   A method of treating water to destroy micro-organisms.

ozone (O$_3$).   A gas that, in the stratosphere, helps protect the earth's surface from ultraviolet radiation from space. Near the surface, or the troposphere, ozone can be created from various sources including vehicle emissions and can be damaging to humans, plants and animals.

ozone depletion potential (ODP).   A value which represents the potential effect of a chemical on stratospheric ozone depletion. The value of 1 ODP is assigned to CFC-11 while some halons have an ODP as high as 10. Substitutes for CFCs, such as HCFCs, can have an ODP of less than 0.05.

ozone hole.   Region first discovered over Antarctica where, in the spring, a significant reduction in stratospheric ozone occurs due to chemical reactions between chlorofluorocarbons (CFCs) and halons and ozone. A so-called ozone hole has also been discovered over the extreme northern latitudes.

FURTHER READING
Somerville (1996).

ozone layer.   See *stratospheric ozone.*

P

Paasche index. A commonly used index that uses weights from the current period. For example, a Paasche price index may be defined as p_1q_1/ p_0q_1 where q_1 is the quantity consumed in the current period, p_0 is the price of the good in the base period and p_1 is the price of the good in the current period. See *Laspeyres index*.

PAC. Acronym for pollution abatement and control.

Paleocene. Geological epoch of time from 66.4 to 57.8 million years ago.

paleoclimatology. The study of past climates through the use of physical samples, such as ice cores.

paleoecology. The study of past ecological systems using the fossil record and associated sedimentary strata.

Paleozoic. Geological era from almost 570 million years to 245 million years ago.

pampas. Grassland region that predominates in Argentina.

panchayat forests. Forests in the Himalayas which are managed by statutory village councils. In Nepal, panchayat forests are often managed by several villages and are managed in the interest of the village communities.

FURTHER READING
Arnold and Campbell (1986).

pandemic. An epidemic that affects various locations simultaneously. The 1919 "Spanish Flu" pandemic simultaneously appeared at various places in the world and was responsible for up to 20 million deaths.

panel data. Data for which observations are available over time, and for individual units of observation. See *cross-sectional data* and *time-series data*.

Pangaea. The name given to a supercontinent that existed from 280 to 180 million years ago.

panspermia.   The hypothesis that life in the universe can be "seeded" from planet to planet via the transmission of micro-organisms in meteors or comets.

parallel evolution.   Evolution of geographically separated, but related, species that results in very similar organisms in terms of appearance or behavior.

parameter.   Fixed values used in a model or optimization problem. Models should be analyzed to determine the sensitivity of the results to values of the parameters.

parasites.   Organisms that live on or in other creatures. The relationship sustains the parasite but often weakens and sometimes kills the host. Many diseases are caused by parasites, including malaria, and have been extremely important in determining evolution in both plants and animals.

Pareto criterion.   A social welfare criterion that finds a change in the state of the world to be welfare improving if, for every individual, the expected utility that results from the change is at least equal to the status quo level of utility, and strictly greater for at least one individual.   See *Pareto improvement.*

Pareto efficiency.   A situation where it is not possible to change the allocation of goods and services to make any individual better off without making someone else worse off.

Pareto improvement.   A change in the state of the world in which the expected utility of every individual in the new state is at least equal to the status quo level of utility, and strictly greater for at least one individual. See *Pareto criterion.*

Pareto optimality.   An allocation of resources or goods in which there exists no other feasible allocation that could improve the utility of one individual without diminishing the utility of at least one other individual.

parthogenesis.   Development of offspring from unfertilized eggs.

partial correlation.   The correlation between two variables accounting for the effects of another variable.

partial equilibrium analysis. Economic analysis which only considers the effects of changes in one sector or market, or part of an economy, without considering the possible feedbacks in the rest of the economy.

participation rate. The proportion of individuals in a defined population who are active in the paid workforce.

particulate loading. The combined weight of particles of a given substance found in a given volume of air or water.

particulate matter. Fine microscopic matter suspended in air or water. Common particulates in the air include smoke and dust. See *suspended solids*.

passive adaptive management. A method of environmental management where managers choose a model and management strategy based on the best available evidence, and then assume that the model and strategies are correct.

FURTHER READING
Hilborn and Walters (1992).

passive gear. Fishing gear that is stationary (such as a gill net). See *active gear*.

passive use value. See *non-use value*.

pasteurization. A method of heat-treating food and drinks to destroy pathogens.

pathogenic organisms. Organisms, usually microbes, that can cause sickness or death, especially in humans.

pathogens. Micro-organisms, including bacteria and viruses, that can be harmful to human health.

pay-as-you-earn (PAYE). A common method of collecting income taxes where the estimated amount owed at the end of the financial year is automatically deducted out of the pay packet of wage and salary earners.

payback period. The period of time that it takes for an investor to repay or cover the initial cost of the investment. The shorter the payback period, everything else equal, the more lucrative is the investment.

PCBs. See *polychlorinated biphenyls.*

peak-load pricing. Pricing that differs depending upon the demand for the good, typically varying directly with the amount demanded. For example, the cost per unit of electricity could vary with the time of the day, with the price increasing at times of peak demand, such as from 17:00–21:00.

peat fuel. Fuel derived from the accumulation of partially decayed organic material where a lack of oxygen has inhibited decomposition.

pecuniary externality. An externality the effects of which are fully accounted for in the price of goods and services such that no distortion exists to the efficient allocation of inputs or outputs. See *externality* and *technological externality.*

pedogenesis. The formation of soils through natural processes.

pedosphere. The portion of the earth consisting of the three layers of soil above the lithosphere.

pelagic species. Fish species commonly found near the surface, such as herring or tuna.

pellagra. A debilitating disease, associated with poverty, that arises from a deficiency of the B-group vitamin, niacin.

per capita income. The total income earned in a region or country in a 12-month period divided by the total number of people in the population.

perennial. Plant whose life cycle goes beyond two years.

perfect competition. A hypothetical benchmark in economics where there exists an identical product; many buyers and sellers, each of whom cannot influence the market price; market participants who all have perfect information about relevant markets; and no barriers to entry and exit of firms.

perfluorocarbons. Greenhouse gases (GHGs) that can have a global warming potential up to 9,000 times greater than carbon dioxide. Perfluorocarbons are a byproduct of aluminum smelting and have been used as a substitute for chlorofluorocarbons (CFCs) in semiconductor manufacture.

performance-based standard. A standard for pollution control that specifies the amount of permissible discharge, but does not dictate the type of technology required to achieve the standard.

performance bonds. See *environmental bonds*.

periodic. A repetitive cycle or change in a variable.

periodic harvesting. Spasmodic harvesting or exploitation of a renewable resource. Periodic harvesting can allow the resource to recover or increase in size so that exploitation can occur at high levels of abundance, thus reducing harvesting costs.

permafrost. Ground that has been frozen for a period of at least two years. See *tundra*.

permanent income hypothesis. The notion that people's consumption is governed by their permanent income, and thus individuals tend to smooth out temporary increases or declines in consumption by saving or borrowing.

permeability. Measure of how well a substance permits liquids or gases to pass through it.

Permian. Geological time period lasting from 286 to 245 million years and which ended with a mass extinction of species.

persistence. Term used to describe substances or processes that linger in the environment. See *persistent organic pollutants*.

persistent organic pollutants (POPs). A variety of pollutants which persist for a long time in the environment and can bio-accumulate, such as DDT or PCBs.

persistent organochlorines (POs). See *chlorinated hydrocarbons*.

PERT. Acronym for pilot emission reduction trading. See *pollution permits*.

perverse incentives. Incentives that lead to undesirable behaviors. For example, in a fishery with no property rights a perverse incentive exists to catch before someone else any fish that can be profitably harvested, regardless of the long-term consequences for the resource.

pesticide resistance. Genetic development, through mutations in pests, that leads to the use of pesticides becoming less effective.

pesticides. Agents, usually chemical, that deter or kill animals (especially insects) that might damage crops or other biological resources.

pH. Abbreviation for potential of hydrogen. A logarithmic measure of the concentration of hydrogen ions in a solution that is used to determine whether a liquid is acidic, neutral or base (alkaline). A pH measure less than 7 is acidic, 7 is neutral and greater than 7 is basic (alkaline). See *logarithmic scale.*

phase diagram. Diagrams which are used to convey the dynamic properties of differential and difference equations.

FURTHER READING
Shone (1997) and Grafton and Sargent (1997).

phenology. Study of the effect on the behavior of animals of periodic changes in the weather and climate.

phenotype. The observable appearance and characteristics of an organism, which are a function of its genetic make-up and the environment.

pheromone. Substances secreted by animals and used to influence the behavior of members of their own species, often in terms of attracting mates. Pheromones are widely used in integrated pest management as an alternative to pesticides.

Phillips curve. An apparent inverse relationship between the rate of increase in wages and the unemployment rate that was first observed by Alban Phillips in a paper published in 1958. The implication of the so-called Phillips curve is that there exists a trade-off between the rate of increase in the price level and unemployment.

phosphates. Sources of phosphorus that occur naturally but which were also found in detergents in the past. Excessive phosphate run-off from fertilizer applications can be a major water pollutant and can accelerate eutrophication of lakes and rivers.

phosphorus. A non-metallic solid element and one of the essential biological nutrients, the availability pf which is a determinant of biological productivity. Phosphorus is a structural component of nucleic acids, phospholipids, and bone in living organisms.

phosphorus cycle.   The movement of phosphorus through soil and water in the environment. See *phosphorus*.

photochemical.   Term used for chemical reactions that arise in the presence of sunlight. See *photochemical smog*.

photochemical smog.   Air pollution is neither smoke nor fog, but is formed by the interaction of hydrocarbons and nitrogen oxides in the presence of sunlight. In large cities with many automobiles, and where air circulation is limited, photochemical smog can be a major problem and health hazard.

photoperiodism.   The seasonal response of plants and animals to changes in the hours of sunlight.

photosynthesis.   A fundamental process of life on earth whereby plants absorb solar radiation and convert it into chemical energy with chlorophyll and, in the process, take in carbon dioxide and water, store carbohydrates and release oxygen.

photovoltaic electricity.   Renewable energy source whereby light from the sun passing through thin layers of silicon can make electrons move between layers and thus act as a battery.

phylogeny.   The evolutionary history of a species, family, order or class.

phylum.   (Plural phyla.) Classification of organisms by classes. See *class*.

physics.   The study of the general principles that govern the physical world.

physiocrats.   A school of thought in eighteenth-century France that promoted free trade and believed that the basis of economic wealth was agriculture.

phytoplankton.   Plankton capable of photosynthesis. See *plankton* and *zooplankton*.

Pigouvian subsidy. See *pollution subsidy*.

Pigouvian tax. Tax named after the economist A.C. Pigou. A Pigouvian tax (commonly a per unit charge) is a tax applied to emissions or pollution and is designed to internalize an environmental externality.

FURTHER READING
Anderson (1994).

pioneer community. First group of plants and animals to first arrive or colonize a site following a major disturbance, such as a forest fire or volcanic eruption.

piosphere. Degraded soils and vegetation surrounding a watering hole or well, usually found in the arid and semi-arid regions of Africa.

placebo effect. An apparent beneficial effect from patients in a clinical trial who receive a non-beneficial substance, but which they believe might be a treatment.

plankton. The aggregate term for a large group of floating, drifting, and sometimes mobile plants and animals found in the water column. See *phytoplankton* and *zooplankton*.

planned economy. An economy where the state or central government, rather than market forces, makes most of the production and pricing decisions. The Soviet Union was a planned economy.

plastics. Materials commonly formed from petroleum and which have a wide variety of uses. Their production, use and disposal can cause significant environmental problems.

plate tectonics. The study of planetary changes in the earth's crust based on the understanding that the continents "float" on molten rock beneath them.

Pleistocene. Geological epoch from around 1.6 million years to 10,000 years before the present, and which coincided with several ice ages.

Pliocene. A geological epoch period from 5.3 million to 1.6 million years ago.

plutonium. A metal formed by nuclear fission of the isotope uranium 238, having a half-life of over 24,000 years.

PM$_{10}$. Particulate matter of a diameter less than 10 micrometers.

PM$_{2.5}$. Particulate matter of a diameter less than 2.5 micrometers.

point source. An identifiable point or location of emissions and discharges, such as a factory stack. See *non-point source.*

Poisson distribution. Named after the French mathematician, Siméon Poisson, it is widely used for predicting the number occurrences of a random event over time and spatially. The cumulative distribution function is defined by

$$\Pr(X=x)=(e^{-\lambda}\lambda^{x})/x!$$

where λ is the mean of the distribution and x! is x factorial.

policy decentralization/centralization. The process of making policy where decision making either takes place within one centralized location within an organization (which can either be a geographical location or a hierarchial level within the organization in the case of policy centralization), or is dispersed throughout an organization (either geographically or among different management levels, or across different decision makers in the case of policy decentralization).

policy failures. Government policies or laws that induce environmental problems, such as subsidies that encourage exploitation of vulnerable resources.

political economy. Term formerly used for the discipline of economics and which today is used to describe the relationship between economic outcomes and policies.

pollutant standards index (PSI). A US index of air quality that combines readings of concentrations of particulate matter, carbon monoxide, ground-level ozone, nitrogen oxide and sulfur dioxide. An index of greater than 100 is considered unhealthy, while an index in excess of 300 is viewed as hazardous.

polluter-pays principle. The idea that those persons responsible for the immediate cause of pollution should also be responsible for paying the costs to society of pollution. One method of implementing the principle is to impose Pigouvian taxes on polluters. See *Pigouvian tax.*

pollution. Any substance that, at the concentrations found in the environment, can have a deleterious affect on the health of individuals, species, habitats and ecosystems.

pollution abatement. The reduction in the level of pollution by the use of end-of-pipe treatment and cleaner technologies, or by the reduction in the output of the product that results in pollution.

pollution allowances. See *pollution permits*.

pollution charge. See *Pigouvian tax*.

pollution credits. See *pollution permits*.

pollution havens. Countries with low environmental standards, or a lack of regulations, that could provide a low-cost place for high-polluting industries to operate.

pollution permits. A method of pollution control where firms are assigned a right to emit a given pollutant in a set period of time, and are allowed to trade this right with other firms in the program. Pollution permits with heterogeneous firms allow firms with high costs of pollution abatement to meet their pollution targets or obligations by purchasing pollution permits from low abatement cost firms. Trading of pollution permits has the potential to reduce the overall cost of achieving reductions in the total level of pollution from an industry.

pollution subsidy (per unit). A per-unit payment for each unit of pollution reduced from a defined level.

polychlorinated biphenyls (PCBs). Organochlorines that have been widely used as an electrical insulator, and in such products as brake linings and paints. PCBs accumulate in fatty tissue and are a known carcinogen in animals as well as being linked to various health complaints in humans.

polychlorinated dibenzodioxins (PCDDs). Commonly called chlorinated dioxins, these are a type of organochlorine. PCDDs bioaccumulate in the environment and are highly toxic.

polymer. A compound formed by a repeated sequence of molecules, such as plastics.

polymorphism. The existence of more than two distinct physical forms of a species, that is not a result of sexual dimorphism. See *dimorphism.*

polynomial equation. An equation of the form
$$a_0 + a_1x + a_2x^2 + \dots + a_nx^n = 0$$
where $a_0 \dots a_n$ are real numbers and a_n does not equal 0. The values of x that solve a polynomial equation are called its roots.

polynomial function. A function of the form
$$P(x) = a_0 + a_1x + a_2x^2 + \dots + a_nx^n$$
where $a_0 \dots a_n$ are real numbers. See *polynomial equation.*

polythene. See *polyethylene.*

polyethylene. A type of plastic that can be recycled and, if disposed of by combustion, generates carbon dioxide and water.

polyvinylchloride (PVC). Polyvinylchloride is a widely used plastic that is made from the highly toxic vinylchloride monomer. PVC does not break down easily and, when burned, can produce toxic compounds.

Pongidae. Family of species that includes chimpanzees and gorillas.

Pontryagin principle. See *maximum principle.*

population. The total number of individual members of a species in a defined geographical area at a particular point in time.

population density. Population per unit of area.

population dynamics. Fluctuations in the population size of a species over time.

population ecology. The study of the interaction of species, genera, or other taxa at the level of populations.

population explosion. 1. The rapid increase in the human population on earth. 2. Any massive increase in a population of a species.

population growth. Net increase in the number of individuals in a defined group over a defined period of time.

population pyramid. Hierarchical ranking of the population of different organisms found in a specific area (grouped by either species or common characteristics), under which those organisms found in a specific rank or row will prey or consume those organisms found in the rank immediately below. These rankings are distinguished by the fact that the biomass and range of the organisms found in the different rows decreases with increasing rank.

Porter hypothesis. The hypothesis that environmental regulations can stimulate innovation and increase productivity by providing incentives for individuals and firms to develop new production processes.

portfolio choice. Term used for the theory and practice of choosing the appropriate set of investments under the assumption that, for a given expected return, investors prefer less risk. See *risk* and *capital asset pricing model.*

positive analysis. See *positive economics.*

positive economics. A term used to describe economic analysis that views the world as it is (rather than as it should be) and, supposedly, is free of value judgements. See *normative economics.*

positive externality. An action which generates benefits for others. These benefits are not taken into account in the decision making of the persons responsible for the action. See *externality* and *negative externality.*

positive feedback. See *feedback.*

posterior distribution. Distribution obtained after an event or following the receipt of new information. See *Bayes theorem.*

post hoc. Latin term meaning "after this" or "following this" in a temporal sense.

post material values. Those human values associated with a society in which material goods are considered unimportant and the primary focus is placed upon human equality, spirituality, and sustainability.

potential energy. The energy which exists in an object by virtue of its position this being the product of its height above the ground, its mass and its acceleration due to gravity.

potential GDP. The total amount of goods and services in an economy that can be produced in a year if all factors of production are fully utilized.

potentially responsible parties (PRPs). Term used in determining liability for US hazardous waste sites. It includes all those persons and firms that may, at one time, have been involved with either the operation or transportation or treatment or disposal of substances at the site.

potential Pareto improvement (PPI). A change, perhaps hypothetical, in the state of the world in which aggregate benefits of the change outweigh the costs. It requires that the potential beneficiaries of the change be able to compensate potential losers so that the expected utility of each individual could, potentially, be at least equal to the status quo levels, and greater than status quo levels for at least one individual.

poverty. A relative and absolute term for the least well-off in a society. Poverty has important environmental consequences as short-term day-to-day survival may prevent the poor from taking a long-term perspective or making investments to improve the quality of the environment.

FURTHER READING
Ray (1998).

power plants. Facilities used to produce energy and, in particular, electricity. Some power plants, such as coal-fired electricity plants, produce sulfur dioxide as a by product, which contributes to acid rain.

ppb. Parts per billion by weight or by volume. See *ppm*.

ppm. Parts per million is a unit of measure used for the concentration of a substance by weight or volume. For example, if a substance is found in the ratio of 4 milligrams per kilogram, it has a concentration of 4 ppm.

prairie. Grasslands found in semi-arid areas of north America, particularly in the mid-west United States and the provinces of Alberta, Saskatchewan and Manitoba in Canada.

Precambrian. A period that encompasses the time when the earth was formed some 4,550 million years ago until 570 million years before the present.

precautionary principle. A principle designed to lead to the sustainable use of the environment and natural resources. It states that actions should be taken to avoid damages in the future even when there exists a lack of full scientific certainty over the potential consequences in the future. The precautionary principle provides a justification for acting today to address possible problems associated with climate change due to anthropogenic emissions of greenhouses gases (GHGs).

FURTHER READING
Perrings (1991).

precipitation. The rain, snow, hail and water vapor that reaches the earth's surface.

predation. The act or process by which predators harvest their prey.

predator. An animal that hunts other animals, its prey, for food.

predator–prey model. A model that explicitly considers the interactions between predators and prey in predicting their population dynamics. See *Lotka–Volterra model.*

FURTHER READING
Grafton and Silve-Echenique (1997).

predetermined variable. Lagged endogenous variables in a simultaneous equation model.

preferences. The underlying wants and needs of individuals.

preferential use doctrine. A bureaucratic means of water allocation in which end uses are given hierarchical rankings that determine the priority of access to water.

present value. The value of returns in the future in terms of current dollars. See *currentavalue (dollars)*
present value of benefits (PVB). The discounted numeric value of future or past benefits.

present value of costs (PVC). The discounted numeric value of future or past costs.

preservation. The practice of environmental management that seeks to ensure that the environment is left, as much as possible, undisturbed and free from anthropogenic influences. See *conservation.*

preservation costs. The costs associated with preserving or maintaining a natural area or wildlife population.

pressure (atmospheric). Commonly measured in millibars and frequently used to predict changes in the weather.

pretreatment standards. Environmental standards on the quality of effluent entering treatment facilities. It may include, for example, restrictions on how much of a certain compound can enter a municipal sewer system.

prey. Animals that are hunted by and fed on by predators.

price ceiling. Regulations that impose a maximum price that can be charged for a good or service.

price discrimination. The selling of the same good at different prices for reasons other than differences in costs.

price elasticity. The percentage change in the quantity demanded (if demand price elasticity) or supplied (if supply price elasticity) following a one percent change in the price. A good is price elastic if a one percent change in the price results in a more than one percent change in the amount demanded or supplied, and is inelastic if a one percent price change leads to a less than one percent change in the amount demanded or supplied.

price floor. See *price support.*

price support. The payment by the government or a state enterprise of a minimum price for a good irrespective of the market price. Price supports most commonly exist for agricultural products.

primary effects. The direct impacts of different policy actions or interventions on the users of a resource stemming from changes in the resource created by the policy actions (for example, improved water quality in a reservoir may lead to increased tourism). See *secondary effects.*

primary forest. A mature forest that has developed with little direct intervention or human activity.

primary fuels.   Sources of energy that can be used in, more or less, the same state as they occur naturally, such as natural gas, coal and oil (and its derivatives).

primary pollutant.   Polluting compounds and   material that are directly discharged into the environment.

primary producer.   1.   In economics, an agent or firm that grows or harvests a resource and converts it into an economic commodity.   2. Green plants that convert sunlight into organic materials.

primary sewage treatment. The first stage, and most basic type, of wastewater treatment in which physical operations are used to remove contaminants.   The most common method is to use a grid chamber and sedimentation tank to remove sediments.   See *secondary sewage treatment*.

primates.   An order of mammals that includes humans, the great apes, lemurs and monkeys.

principal components analysis.   A   statistical   approach   to   help overcome problems of multicollinearity that tries to extract much of the variation present in the independent variables in a regression.   See *multicollinearity*.

FURTHER READING
Greene (1997).

prions.   Proteins that have been linked to some diseases, such as "mad cow disease".   See *bovine spongiform encephalopathy*.

prior appropriation doctrine.   The notion that the first user of a resource, such as a river or a stream, has the uninterrupted right to the resource equal to the amount initially withdrawn or used by the individual.

prisoner's dilemma.   In game theory, a model in which two prisoners (agents) are given the opportunity to confess to a crime or deny a crime with the understanding that the outcome of this decision depends on the decision of the other prisoner.   If both deny the crime, both face a lenient penalty, if one confesses and the other denies then the one who confesses faces the most lenient penalty while the one who denies faces the most severe penalty, and if both confess both face intermediate penalties.   The prisoner's dilemma demonstrates why individuals may choose to engage in non-cooperative behavior, even when co-operative behavior has a higher combined payoff, and is used to describe more general games and real-world situations where the same incentives hold.

private good. A good where the owner can exclude others from using it, its use being rivalrous in the sense that it precludes others from using it simultaneously. An example of a private good is an automobile.

private property rights. See *private rights*.

private rights. A property-rights regime where ownership of goods and resources rests with individuals or firms.

FURTHER READING:
Grafton, Squires and Fox (2000).

privatization. The divestiture of ownership or control of assets, or the right to use resources, by the state to individuals or groups of individuals. Privatization has been undertaken to increase revenues for the state by the sale of state assets, and also to increase economic efficiency.

probability. The chance or likelihood of an event occurring, defined by $Pr(x)$ where x is the event. If the event will occur with certainty, then $Pr(x) = 1$ and if the event will never occur, then $Pr(x) = 0$.

probability density function (PDF). See *probability distribution function*.

probability distribution function. Sometimes called the probability density function, it is the function that assigns a probability of occurrence for each possible value of a discrete random variable. Thus for a random variable X, the probability function can be used to obtain $Pr(X=x)$ where x is any value that the random variable can obtain.

probit analysis. A method of regression analysis with a limited dependent variable (usually binary) that is commonly used for discrete choice analysis. The probit model assumes a functional form for the regression in which $y_i^* = \beta x + u_i$, where y_i^* is unobservable, but can be deduced by assuming that an observable dummy, y, equals 1 if $y_i^* > 0$ and 0 otherwise. In the probit analysis, u_i is a random variable with a cumulative normal distribution.

FURTHER READING
Greene (1997).

process and production methods (PPMs).   Term used to describe how products are produced, which may have important trade implications. The term generally refers to non-product-related PPMs where there is very little difference in final products, despite differences in the production process. Trade law under the World Trade Organization (WTO) prohibits discrimination of products on the basis of non-product-related PPMs. Discrimination on product-related PPMs, where there exist significant differences in the final products, such as for genetically modified potatoes versus organically grown potatoes, is permitted under trade law.

FURTHER READING
International Institute for Sustainable Development (2000).

producer surplus.   Returns over and above those necessary to compensate a firm to supply a given quantity of a good or service.

production costs.   The direct costs associated with the production or manufacture of a good, asset or output. See *variable costs*.

production function.   A function where the dependent variable is output, and the independent variables are inputs in the production process.

production possibilities frontier (PPF).   Also known as the production possibilities curve, this represents a mapping of the maximum quantities of the different possible combinations of goods and services that can be produced in an economy in a given period of time, using the current resource base and technology.

productivity.   A measure of total output divided by a measure of the inputs to produce the output. A common productivity measure is labor productivity which divides the value of output by the number of worker hours or days taken to produce it.

productivity–diversity relationships.   The relationship between the number of species in an ecosystem and the rate of growth of one or more resources in that system that is subject to biological competition.

FURTHER READING
Abrams (1995).

product substitution. The degree to which one product will replace or be replaced in consumption by another product following a change in its price.

profit. The difference between total revenue and total cost. See *normal profit.*

profit function. A function in terms of output and input prices that defines the maximum level of profit for a firm at any given set of prices.

profit maximization. A common economic assumption about firm behavior that supposes firms optimize by maximizing their profits (that is, revenues less all costs of production).

progressive tax. A tax which takes proportionately (as well as absolutely) more from people with higher incomes than with lower incomes. Most income taxes are, to some extent, progressive in that the highest marginal rates of taxation are imposed at higher levels of income.

prokaryote. Organisms that do not have nuclei in their cells. See *eukaryote.*

property rights. The *de jure* and *de facto* rights of individuals or groups of individuals to use or benefit from resources, goods and services.

FURTHER READING
Bromley (1991) and Devlin and Grafton (1998).

property-rights regime. A way of describing or grouping property rights with similar characteristics. A commonly observed property regime is private rights.

prospect theory. Psychological theory that states, among other things, that individuals value losses from a given state or status quo much more than they do gains of equivalent magnitude.

protandry. The change in gender of an organism during its life from a male to a female.

protective/recovery measures. Measures to protect a species, as required under the United States Endangered Species Act. Protective measures include restrictions on activities that may harm a species, while recovery measures give the United States Federal Government the authority to improve the condition of a listed species.

proteoid.   Plants characterized by thickened leaves.

Proterozoic era.   A geological era from 2,500 to 570 million years ago.

protozoa.   Single-celled organisms that are responsible for a variety of human diseases, including leishmaniasis.

proved reserves.   Estimated reserves of a non-renewable resource that can be extracted with a reasonable degree of certainty.

psychrosphere.   Part of the sea or ocean where the temperature is low and varies little over the seasons.

public choice theory.   Economic theory concerned with how the government and the state actually make decisions.   It considers political self interest to be an explanatory factor.

public debt.   See *national debt.*

public good.   A pure public good is non-exclusive (anyone can use it) and non-rivalrous (the number of users does not affect the benefits derived from its consumption). Public goods may be publicly provided (such as national defense) or may be privately provided (such as a lighthouse).

public investment.   The aggregate expenditures by a government on public goods.

pulp.   A product formed from wood chips that is used for making paper. Some of the chemical processes to derive pulp can generate considerable water and air pollution.

pulp and paper mill.   A factory that converts wood chips into pulp and paper. Pulp and paper mills can be major point sources of both air and water pollution.

pulpwood.   Wood that, upon harvesting, is directly transformed into pulp and paper.

pulse fishing.   See *periodic harvesting.*

punctuated equilibrium. The notion that the evolution of life is characterized by long periods of relative stability punctuated by comparatively short periods of major evolutionary change.

purchasing power parity (PPP). A method of comparing the value of currencies across countries using a suitable basket of goods and services. For example, if the basket of goods costs $A400.00 in Australia and $US200.00 in the United States then the purchasing power parity is $A2 for $US1. The PPP may differ from the actual exchange rate.

pure compensation. Term used to describe population growth models where declines in the population are compensated for by increases in the population growth rate.

pure competition. See *perfect competition*.

pure existence value. The value ascribed to a good solely for its existence and divorced from any possible *in situ* use value it might have.

FURTHER READING
McConnell (1983) and Krutilla and Fisher (1975).

purse-seiner. A fishing vessel that employs a net that encircles schools of fish and is then closed from below, trapping the fish. This method   of fishing is employed to catch tuna, herring and other pelagic species, as well as salmonids.

PVC. See *polyvinylchloride*.

pyrotoxins. Toxins produced by fire.

Q

QED.   See *quod erat demonstrandum.*

quadrats/quadrant sampling.   Sampling or measurements done within square-sided plots (e.g., looking at the density of different plant species by the amount of area covered by the plants).

quality of life.   A broad and holistic measure of the well-being of individuals within a society that includes consideration of health and health care, education, crime, environmental quality, and so on. See *genuine progress index.*

quality of title.   A property-right characteristic that refers to the *de jure* rights of the holder.

quantified emissions limitations and reductions objectives (QELROs).   The emission targets for greenhouse gases (GHGs) that Annex B countries in the Kyoto Protocol have committed themselves to meet. See *Kyoto Protocol.*

quasi-option value.   The welfare gain that comes from delaying a decision to use a natural resource when there are uncertainties about the payoff of alternative uses of the resource, and when at least one of the uses is irreversible.

FURTHER READING
Arrow and Fisher (1974).

quasi-rent.   Earnings or returns to a fixed factor of production that is temporarily fixed in supply. Quasi-rent is the surplus over and above the variable costs of production but may still be less than the fixed costs and, thus, positive quasi-rents do not necessarily imply positive profits.

Quaternary period.   Geological period of time from 1.6 million years ago to the present.

quid pro quo.   Latin term for a mutual agreement to exchange. The term generally refers to the implicit understanding that a favor granted from one person to another should be reciprocated.

quod erat demonstrandum (QED).   Latin term meaning "that which was to be proved". The term is commonly used at the conclusion of a proof.

quota.   A regulated amount of a good that can be produced, imported or exported by a firm or country.

R

race to fish. Term used in so-called derby fisheries where the total harvest is controlled, but the individual harvests of fishers are not. Consequently, fishers compete or "race to fish" among themselves to harvest as much as they profitably can before the fishing season closes. See *derby fishery*.

rad. Unit of measure of absorbed ionizing radiation.

radiation absorption. The accumulation of radioactive waves within animal tissues and plant matter from exposure to radioactive sources.

radiation budget. An accounting procedure that tracks the amount of radiation that reaches and leaves the earth's surface. The budget consists of sub-budgets that account for reflection and absorption by the earth's atmosphere. Changes in the proportion of greenhouse gases in the atmosphere may alter the earth's radiation budget causing global warming.

radiation feedback. See *radiative forcing*.

radiation reflection. See *albedo*.

radiation sickness. Sickness caused by exposure to radiation that includes nausea and diarrhea and, in the long run, lowered immunity and increased rate of cancers.

radiative forcing. Also known as climate forcing, this refers to the perturbation to the energy balance of the earth's atmosphere following a change in the system, such as an increased concentration of carbon dioxide. A positive radiative forcing (due to increased concentrations of atmospheric carbon dioxide) tends to increase mean surface temperatures.

radical. A chemical compound in which there is a free electron, thus making the compound highly reactive.

radioactive waste. Waste material from nuclear fission. High-level radioactive waste, such as spent fuel rods in nuclear reactors, present a particular environmental hazard. Various proposals have been made for the long-term storage of high-level radioactive waste in various countries, but have been met with widespread concern regarding the potential leakage of the material.

radiometric age determination. The dating of material using radio-active isotopes, such as Carbon 14.

radionuclides. Isotopes of different atoms, from both natural and man-made sources, that decay and release ionizing radiation that can be hazardous to human health.

radon. Naturally occurring gas produced from radium, which can seep into the basements of buildings and, thus, create a health hazard.

radon-222. A radioactive gas that can pose a severe risk to human health, and that has been identified as a hazardous pollutant.

rainforest. Mainly evergreen forests where there is usually an abundance of precipitation throughout the year. Tropical rainforests are very important ecosystems that include many of the world's terrestrial species. Rainforests also exist in subtropical areas, such as southern Queensland in Australia, and in temperate areas such as the USA, the Canadian Pacific west coast and the west coast of the South Island of New Zealand.

FURTHER READING
Parks (1992).

rain shadow. Dry or arid area on the leeward side of mountains.

Ramsey growth model. Model of economic growth developed by the English economist Frank Ramsey, which includes the specification of consumer behavior.

FURTHER READING
Barro and Sala-i-Martin (1995).

random effects. Statistical term that refers to a model where group effects on individual units of observations arise through a group specific disturbance term. See *fixed effects*.

FURTHER READING
Greene (1997).

random parameters model. Econometric models in which the values of the parameters do not stay fixed, but change due to one of several reasons including a change in regimes or processes, or because the parameters depend upon other variables, or because the parameters are simply random in nature.

random sample. Sample of observations selected from a population where each observation has an equal chance of being selected.

random shocks. Unexpected and non-deterministic events, the occurrence of which cannot be predicted in advance.

random utility model (RUM). A model of consumer choice in which the consumer is assumed to choose among goods or activities in order to maximize their "random utility", and which consists of a deterministic core plus a good-specific random term.

FURTHER READING
Louviere (1996).

random variable. A variable whose value is determined by a particular probability distribution.

random walk. The movement or motion of a variable over time that moves in discrete jumps from one point in time to another at a fixed probability. A variable that follows a random walk may have an apparent tendency to move in a particular direction, especially if there exists an absorbing barrier that may prevent it moving past a particular point.

range. The difference between the highest and lowest value in a sample of a population.

range management. The regulation of grazing and other activities on open land.

rapid rural appraisal (RRA). A multidisciplinary assessment method in which teams of researchers gather and present information concerning site-specific farming conditions, techniques, and outcomes. RRA is intended to be an informal tool to evaluate farming impacts on the environment, and also a means for fostering co-operation between local people and researchers.

raster.   The representation of spatial or geographic data in a grid.

rate of return.   The net returns from an enterprise as a ratio of the total amount invested.

rate of time preference.   The preference of consumption today relative to the future. A higher rate of time preference coincides with a higher discount rate. See *discount rate.*

rational agents.   Individuals or other economic agents who act in a manner that will maximize their own well-being.

rational expectations.   The hypothesis that market participants use all the available information to predict the future and that, on average, they are correct, barring non-systematic or random errors. Rational expectations led to a questioning of the ability of governments to systematically trick or fool consumers or firms through their policies. See *adaptive expectations* and *Lucas critique.*

FURTHER READING
Sheffrin (1983).

Rawlsian justice.   The notion expounded by John Rawls that decisions should be taken so as to consider the least fortunate or well-off. A Rawlsian social welfare function maximizes the minimum level of utility and has been applied in studies on intergenerational equity.

FURTHER READING
Rawls (1971).

reaction curves.   Also known as reaction functions. The mathematical expression of a participant's set of preferred actions within a game that may, or may not, be based on the actions of other participants.

real GDP.   The value of goods and services produced in an economy in a year measured in terms of real prices, or relative to a base-year index.

real interest rate.   The current interest rate less the current rate of inflation.

real numbers.   All numbers consisting of integers (... -1, 0, 1, 2 ...), ratios of integers and functions of integers.

real option pricing. The method under which the opportunities for a firm to invest in real assets (new technology, plants, or equipment) are valued.

real value. The inflation-free value of a good or service, adjusted to represent the exchange value of that good or service for a representative basket of market goods.

real wage. A measure of earnings by workers divided by an index of the price level.

receptor locations. Specific sites that incur the environmental impacts of pollution.

recession. Term used to describe a small decline or slowdown in the level of economic activity. In some countries, three successive quarter (3-monthly) declines in the gross domestic product is officially defined as a recession.

recharge rate. Rate at which water enters an aquifer.

reciprocity. The mutual assistance provided between individuals and between groups of individuals. Reciprocity helps explain the success of human beings, it being an important component of the collaboration and co-ordination that exists between individuals in a society.

RECLAIM. See *Regional Clear Air Incentives Market.*

reclamation. The process of restoring an environment to a usable state. Reclamation commonly occurs after mining.

recoverable reserves. That portion of the mineral body or deposit still unmined but which can be economically extracted.

recovery plans. Plans or strategies designed to restore an ecosystem or environment to a desirable state, such as after a mining operation.

recreational fisheries. See *sport fisheries.*

recreational site choice models. Models of consumer choice that reflect the consumer's decision to visit one or more multiple recreational destinations. Recreational site choice models provide a means to estimate the non-market values of outdoor recreation sites, as well as environmental amenities at these sites.

recreation benefits.   The benefits from an aspect of the environment associated with recreation. Recreation benefits (such as walking in the forest) are often non-market values but can be estimated through a variety of techniques such as the contingent valuation methods and travel cost methods.

recruitment.   The weight or number of a species that, upon reaching a certain level of maturity, is considered to be part of the biomass or resource stock.

recruitment overfishing.   Fishing that reduces the number of reproductive fish in a population that, in turn, reduces future recruitment and the future population.

recyclables.   Any material that, after its use, can be used as an input in a production process, such as glass and steel.

recycling.   Process of collecting and reprocessing used materials. Curbside or blue-box recycling programs are widespread in many rich countries.

recycling ratio.   The amount of a good recovered through the recycling process expressed as a percentage of the amount entering the process.

red data species.   Rare and endangered species listed by the International Union for the Conservation of Nature. See *International Union for the Conservation of Nature*.

red queen hypothesis.   The notion that species need to continuously evolve to survive.

red tide.   An algal bloom that occurs in coastal waters and which can produce toxins that can be fatal to marine species and humans who eat contaminated animals, especially shellfish. The phenomenon can occur naturally but is also enhanced by the run-off of phosphates into the sea. See *algal bloom*.

reduced form.   A simultaneous equation model rewritten such that the endogenous variables are a function of the undetermined parameters, the disturbance term and lagged endogenous variables.

reductio ad absurdum. Latin term meaning "a reduction to the absurd". For example, if one glass of red wine at dinner helps reduce the risk of heart disease then the belief that drinking a cask of red wine every day eliminates the chance of a heart attack altogether is *reductio ad absurdum.*

reduction. A chemical reaction that involves the loss of oxygen, or the addition of hydrogen, to atoms or molecules.

reference concentration (RfC). Term used by the US Environmental Protection Agency that estimates the level below which daily exposure is unlikely to lead to deleterious health effects in the human population.

referendum. A question format in contingent valuation surveys in which the respondent is asked to respond "yes" or "no" to a hypothetical tradeoff between some amount of an environmental good or service and something else of value (especially money). The referendum format is considered by many to be superior to other question formats, but may require strong assumptions about the functional form of utility in order to derive from it willingness to pay measures.

FURTHER READING
McConnell (1990).

reforestation. The restocking or regeneration of trees and forest after the land has been harvested for trees. Reforestation can be both a natural process (from seeds) or an anthropogenic process (from planting seedlings).

refund system. See *deposit-refund system.*

refuse-derived fuel. Energy derived from the combustion of waste material.

reg. An arid landscape characterized by large areas of gravel.

regeneration. The regrowth of a tree crop through natural or artificial means. See *reforestation.*

Regional Clean Air Incentives Market (RECLAIM). An air pollution permit system in southern California that controls emissions of nitrogen oxides and sulfur dioxides.

regression. The tendency of offspring of parents who are above or below the mean in a particular characteristic (such as height) to revert or regress to a value closer to the mean of the population.

regression analysis. A statistical analysis whereby a relationship between a dependent variable against one or more independent variables is estimated using data. For example, the relationship between the rate of savings (dependent variable) and the level of income (independent variable) can be estimated in a savings–income model. See *ordinary least squares (OLS)*.

regression to the mean. See *regression*.

regressive tax. Tax that imposes a proportionately greater burden on low-income households than on high-income households. Some green taxes, such as a carbon tax, may be regressive.

relative abundance. The frequency with which different organisms or resources are found in different ecosystems.

relative humidity. The proportion of water in the air relative to the maximum that the air is capable of holding at the current temperature.

relative risk. The ratio of the incidence of a complaint, malady or illness to those exposed to a risk factor, relative to the incidence among those not exposed to the risk factor.

remedial actions. Environmental clean-up activities designed to return a degraded site or resource to its original condition.

remediation. Process and practices associated with the restoration of habitats and ecosystems to achieve desired outcomes. Remediation is commonly practised at waste sites and old industrial sites so that the land can be used for other purposes.

remote sensing. A method of obtaining physical data from a distance. Many satellites use remote sensing to provide data on various phenomena, including weather.

renewable energy. An energy source that is undepleted through use or withdrawal, such as the conversion of solar radiation into electricity through photovoltaic cells.

renewable resource. A natural resource, like a fishery, that has the potential to be self-perpetuating.

rent. See *economic rent.*

rent dissipation. Term given for the over-use of scarce resources in the harvesting of a resource or the production of a good. Rent dissipation is common in open-access fisheries where the "race to fish" encourages the over-use of inputs and, thus, an overall reduction in returns from the resource.

rent gradient. Term for the decline in the rent paid for land the further its distance from a central core.

rent seeking. Term used to refer to activities and lobbying by individuals or firms to acquire scarce property rights that can be assigned by a central authority. The term was first used with respect to the activities of importers who wished to have access to import licenses or quotas.

repeated games. Games in which the participants interact more than once and which allow for the possibility of participants modifying their strategy in response to observing the choices of others.

replacement rate. The fertility rate that is necessary to maintain a population of fixed size such that is neither growing nor shrinking in size.

replicate. The process by which some organisms duplicate themselves. Bacteria reproduce by replication.

reproduce. The process by which organisms produce offspring.

reproducible capital. Capital, usually human-made, that can be reproduced exactly; for example, machinery and buildings.

reproductive success. The proportion of offspring that are alive after a given period of time.

required compensation locus. Government policies to manage risk often impose costs on certain individuals, including changes in net income and direct utility costs. The required compensation locus (RCL) may be an infinite number of pairs of state-dependent compensations that would exactly, and at the least cost, return the individual to his or her original, status quo, expected utility.

FURTHER READING
Hazilla and Kopp (1990).

res communes. Latin term for community property rights. See *community rights*.

reserves-to-use ratio. The volume of reserves of a non-renewable resource divided by the current consumption of the resource. The lower the ratio, the sooner the resource will, theoretically, be depleted.

residual. 1. The amount of materials, energy or radioactivity introduced into the process of production or consumption that remains unused, and is returned to the environment. 2. A statistical term for the difference between the estimated value of a variable and its observed value.

residual claimant. The party entitled to whatever profits or returns are left after all factors of production, such as labor, have been paid.

resilience. A term used to describe the ability of natural populations to return to their previous state following a shock.

FURTHER READING:
Holling (1973).

res nullius. Latin term that refers to natural resources which have free or open access and are owned and managed by no one. Many fisheries in the high seas outside the jurisdiction of coastal states may be described as res nullius. See *commons, tragedy of.*

resource allocation. The distribution of goods and assets among individuals.

resource base. The natural resources, including land and water, available to sustain a community or society.

resource recovery maximization. The attempt to recycle and reutilize as much of the waste generated by industrial activity as possible before discarding it into refuse facilities.

resource rent. Also known as scarcity rent, this is the economic surplus that arises wherever supply is constrained such that a marginal increase in supply is prevented, and where such an increase would be profitable.

resource rent tax.   A tax imposed on the extraction of non-renewable resources based upon the cash flow from the operations.

FURTHER READING
Garnaut and Ross (1975).

respiration.   A process by which organisms convert food into carbon dioxide and water and, in the process, release energy.

response bias.   The bias that arises when the information supplied by a respondent differs from reality. Various techniques can be used, including asking the same question in different ways, to reduce response bias.

response rate.   The ratio of people who respond to a survey to the number of people asked to participate.

res privatae.   Latin term for private property rights. See *private rights*.

res publicae.   Latin term for state property rights. See *state rights*.

restoid.   Vegetation having a low, bushy form similar to that of the genus *Restio* found in Africa.

restoration costs.   Expenditures associated with helping to restore an environment to a previous state.

retroactive liability.   A legal rule that makes it possible for individuals, firms or even governments to be held legally responsible for their actions in the past even if those actions were, at the time, legal. See *liability*.

retrofitting.   The conversion of existing equipment or materials to meet pollution or emission standards. For example, the attachment of "scrubbers" to smoke stacks to reduce sulfur dioxide emissions is a form of retrofitting.

retrovirus.   Viruses, such as the virus responsible for AIDS, that can mutate quickly and use the cells of their host to make DNA copies of their RNA.

revealed preference.   An approach developed by the Nobel laureate in Economics, Paul Samuelson, that is used to identify the underlying preferences, and thus demands of individuals, based upon the choices each reveals in their consumption. Thus, if a bundle of goods A is bought when another bundle B is available and affordable, then bundle A is revealed to be preferred to bundle B.

reverse osmosis.   A cost-effective method of filtering water by which untreated water is forced at pressure through a membrane that traps compounds, parasites and other undesirable matter.

reversibility.   See *irreversibility*.

Rhine Convention.   Name given to a series of conventions on the use of the Rhine river. The convention signed in Bonn in 1976 regulated the discharge of specified chemicals into the river by France, West Germany, Switzerland, The Netherlands and Luxembourg.

Rhine spill.   A 1986 discharge of pesticides into the Rhine river that originated from a chemical factory in Basle, Switzerland.

rhizomatous perennials.   Herbaceous plants characterized by the presence of a central enlarged root which produces the stem and leaves.

ribonucleic acid (RNA).   An organic code used in the transfer of genetic information.

Ricardian rent.   Also known as differential rent, this is the economic surplus that arises from the diminishing returns from using successive units of a factor of production, given that other factors of production are fixed in supply. Those units of the variable factor that generate a value of their marginal product that exceeds their cost obtain a Ricardian rent.

Richter scale.   A logarithmic scale for measuring the intensity of earthquakes and tremors from seismic waves. The scale ranges from 1 to 9 where a unit increase in the scale corresponds to a 10-fold increase in the intensity of the seismic waves. See *logarithmic scale.*

Ricker curve.   See *stock-recruitment model*.

rickets.   A disease that affects bone formation,arising from a lack of vitamin D due to insufficient sunlight or from an inability to absorb calcium. The disease is a symptom of poverty and was widespread in northern countries when children were employed inside factories or mines.

rill erosion.   Erosion caused by small rivulets and streams.

Rio plus 10.   The follow-up conference to the United Nations on Environment and Development that was held in Rio de Janeiro in 1992. Another conferenc is scheduled to take place in Durban, South Africa, in 2002.

Rio Summit.   See *United Nations Conference on Environment and Development.*

riparian.   A term meaning near or adjacent to streams, rivers, lakes or wetlands.

riparian rights.   The property rights associated with the access to, and use of, streams, lakes and rivers.

riparian water doctrine.   A system of property rights in which access to water (from a natural water course) requires ownership, or other tenure, of land adjacent to that water.

risk.   Term for a set of circumstance where probabilities exist for all potential events or outcomes. See *uncertainty.*

risk assessment.   Evaluation of the risks posed to health and/or the environment from a given activity.

risk averse.   Term used to describe an individual who prefers a certain outcome with a given return than an uncertain outcome with the same expected return.

risk–benefit analysis.   A type of risk assessment whereby the risks or costs associated with an activity are compared to the potential benefits. For example, breast screening through mammograms imposes some health risks but the benefits in early detection of cancers may, for older women, outweigh the possible health costs.

risk loving.   Term used to describe an individual who will prefer an uncertain outcome with a given expected return, over a certain outcome with the same return.

risk management. An approach to decision making that explicitly considers the risks associated with decisions and investments.

risk neutral. Term used to describe an individual who is indifferent concerning a certain outcome with a given return and an uncertain outcome with the same expected return.

risk premium. The extra return required to induce an individual to hold a riskier asset (returns that exhibit greater variability).

rivalrous. Term used to describe a resource or asset where if someone increases his or her use, less is available to others. All common-pool resources, such as a fishery, are rivalrous in use. See *common-pool resource*.

river basin. The geographic area that contains all the drainage waters of a specific river.

river blindness. See *schistosomiasis*.

river corridor. The land bordering a river channel that is often important in terms of wildlife habitat and species diversity.

RNA. See *ribonucleic acid*.

roadless areas. 1. Areas in which no roads have been built. 2. A term used in the USA to designate National Forest land that falls under special management rules.

robustness. See *resilience*.

Roentgen. A unit measure of absorbed radiation from X or gamma rays that was named after Wilhelm Roentgen.

roots. 1. The part of plants that enables them to uptake nutrients and water from soils. 2. The values that solve a polynomial equation. See *polynomial equation*.

rotation. The process of cutting and replanting a stand of timber.

rotation age. The age at which a forest is scheduled to be harvested. Rotation age can refer to the age of biological maturity, biologically maximal productivity, or the economically optimal age for harvest. See *optimal rotation.*

Rotterdam Convention. The Rotterdam Convention on the Prior Informed Consent Procedures for Certain Hazardous Chemicals and Pesticides in International Trade provides a set of procedures to ensure that importing countries are adequately informed about potentially hazardous materials that are internationally traded.

roundwood. Merchantable sections of tree stems free from branches, and sometimes free from bark, that may include logs, posts, pilings and bolts.

royalty. A payment made to the owner of a natural resource for exploiting the resource, and that is based on the value of the resource extracted.

Roy's identity. A mathematical result used to obtain the demands for a good from an indirect utility function, and that is the negative of the ratio of the partial derivative of the indirect utility function with respect to the good's price to the partial derivative of the indirect utility function with respect to the consumer's income.

r-squared (R^2). See *coefficient of determination.*

r-strategists. Animals that ensure their reproductive success by producing large numbers of offspring, such as insects.

rule of capture. The norm, and often the law, that allows a person who captured or harvested a resource to enjoy its benefits. Rule of capture applies in many renewable resources, such as fisheries.

run-off. The water that is transported across the land and deposited into water bodies such as rivers and lakes. Run-off from urban centers may also include a variety of contaminants, including oils and particulates.

runs test. Test used to detect serial correlation that is often used in time-series analysis. The test compares the sequences of positive and negative residuals with their expected number.

S

saddle-point. A point on a multivariable function that is neither a maximum nor a minimum, but is where the value of the function does not change with small changes in the values of the choice variables.

"Safe Harbor" policy. An agreement in the USA under which non-federal landowners can restore, enhance, or maintain habitats for endangered species with the assurance that additional restrictions will not be placed upon them if endangered species either appear on the land or populations increase.

safe minimum standard. An approach to resource management that tries to reduce the probability of potentially high cost or losses by ensuring the level of exploitation is such that the population does not fall below a safe minimum standard. See *precautionary principle.*

FURTHER READING
Ciriacy-Wantrup (1952).

Sahara desert. The desert region in north Africa that stretches from the Atlantic ocean to the Red Sea.

Sahel. An arid area in western Africa, south of the Sahara desert, which encompasses (from west to east) Senegal, Mauretania, Mali, Burkina Faso and Chad.

sales tax. Tax levied as a percentage of the retail sales price of a good or service.

salinity. The proportion of dissolved salts in water. Water with salinity levels in excess of 500 parts per million (ppm) is unsuitable for domestic use and salinity levels in excess of 2,000 ppm makes the water unsuitable for irrigation purposes. Excess salinity in soils can be a major agricultural problem, especially in poorly drained soils. A major cause of excess salinity in soils is irrigation that can contribute to the rise in the ground-water table and lead to excessive evaporation of water on the surface, such that salt levels increase over time.

salinization. The accumulation of mineral salts in soil over time from the evaporation of water. A problem commonly found in irrigated soils in more arid conditions.

salmonids.   Collective term for salmon species, including trout.

salt marshes.   Marshy areas influenced by ocean waters.

saltwater intrusion.   The movement of seawater into underground freshwater aquifers, often caused by pumping of water from the aquifers.

salvage value.   The value of an asset at the end of a particular period of time, usually its estimated working life.

sample.   A selection of members or elements from a larger population. A great deal of statistical analysis involves tests on a sample to infer the characteristics of the population.

sample selection bias.   The error that arises from making inferences about a population from a sample or samples that are not representative of the population.

sampling error.   The difference between the estimator obtained from a sample and the estimator obtained from a population.

Samuelson conditions for public goods.   A condition that ensures a Pareto-efficient outcome in the provision of a pure public good that requires the sum of the marginal rates of substitution for the private and public goods (absolute value of the slope of the consumers' indifference curves) for all individuals to equal the marginal rate of transformation (absolute value of the slope of the production possibilities frontier).

FURTHER READING
Cornes and Sandler (1996).

sanitary landfill.   A municipal or industrial waste or dump that meets defined standards to avoid leaching and contamination of soils and groundwater.

saprophyte.   Organisms that live off the dead, decaying or non-living parts of other organisms.

satellite accounts.   Financial accounts, especially in systems of national accounts, that consider economic and physical stocks and flows (e.g., the value of oil reserves) that relate to traditional accounts (e.g., gross domestic product), but are not considered part of the system of core accounts.

satisficing. Term coined by the economist Herbert Simon, denoting that individuals and firms seek to achieve certain levels or goals rather than the highest possible values. For example, firms may wish to achieve a certain profit level that is less than the maximum possible, after which they may seek to achieve other goals.

savanna. A biome characterized by little rainfall and grasslands with scattered trees, that is found in tropical and subtropical regions.

sawntimber. Wood that has been harvested and sawn either into lumber or into large pieces of timber with square sides.

Say's law. A law, attributed to the nineteenth-century economist Jean-Baptiste Say, commonly defined as the idea that supply creates its own demand.

scale efficiency. Production at the level of output that maximizes profits. See *allocative efficiency, economic efficiency* and *technical efficiency*.

scarcity. A measure of the relative or absolute abundance of a good or resource. For example, a widely used scarcity measure is the ratio of proven reserves to current annual use of a non-renewable resource – a measure that may actually increase over time due to the discovery of new reserves or a decline in consumption due to rising prices.

scarcity rent. See *resource rent*.

Schaefer–Gordon model. See *Gordon–Schaefer model*.

Schaefer model. A surplus production model of a fishery where the growth in the biomass is defined by a logistic growth function less the harvest of fish. See *logistic growth* and *surplus production models*.

schistosomiasis. A sometimes fatal group of water-borne diseases caused by parasitic worms. The disease is contracted from larvae that can penetrate the skin of persons in water and is common in Africa, Asia and South America. A common form of the disease, bilharziasis, is endemic in most parts of Africa and has increased with the building of dams and the creation of large bodies of water, which have increased the population of snails that serve as hosts in a stage of the life-cycle of the parasites.

scientific method. An approach to understanding and explaining the world around us that requires all ideas to be questioned and tested. The method requires that ideas, theories or hypotheses be formed in a dynamic interaction of observations, tests and experiments.

Scitovsky compensation criterion. One social state is preferred to another when, using the Kaldor criterion, it always leads to that state being preferred and when the Kaldor criterion does not lead to contradictory results. This is equivalent to the Pareto criterion if there are no restrictions on reaching any social state from any other social state.

sclerophylous. The condition by which evergreen shrubs and trees survive prolonged droughts by producing tough, thick-skinned leaves.

scrubbers. Attachments to smoke stacks where emissions from the combustion of coal can be treated with limestone, or lime-rich solutions, to reduce sulfur dioxide emissions.

scurvy. A disease due to a dietary deficiency of vitamin C. Its symptoms include painful joints and membranes, and skin eruptions.

secondary effects. The indirect impacts created by changes in the use of a resource on those activities that do not directly involve the resource itself. For example, improved water quality in a reservoir may lead to improved tourism which in turn may increase the demand for seasonal help. The increased demand for help would be a secondary effect. See *primary effects*.

secondary forest. A regenerated forest that has grown after harvesting, or following a major disturbance, such as a forest fire.

secondary pollutants. Pollutants that are formed from subsequent chemical reactions once primary pollutants are discharged into the environment (such as the production of ozone by the interaction between nitrogen oxides and volatile organic compounds in the atmosphere).

secondary sewage treatment. A second step in waste treatment in which micro-organisms consume most of the solids remaining after primary treatment. A common method of secondary treatment is to use a trickling filter that sprays wastewater over rock beds coated with micro-organisms. See *primary sewage treatment*.

second best. See *theory of the second best*.

second fundamental theorem of welfare economics. An economic theorem that states that any Pareto-efficient outcome can be obtained from a perfectly competitive market, under a certain set of conditions and with appropriate lump sum transfers or initial wealth distribution between individuals. See *Pareto efficiency*.

second growth. Forest that has regenerated following a harvest, or a natural disaster such as a forest fire, and which has yet to reach maturity.

second law of thermodynamics. Commonly called the entropy law. It states that the amount of unavailable energy (entropy) in a closed system (where there is no exchange of matter) continuously and irreversibly increases over time. As a result, in a closed system, low entropy forms of energy decrease over time and reduce the available energy for work. See *entropy* and *law of conservation of energy*.

FURTHER READING:
Georgescu-Roegen (1973).

second-order conditions. Conditions for an optimum that are obtained from the second-order partial derivatives of an optimization problem.

FURTHER READING
Grafton and Sargent (1997).

sedentary species. Species that remain in one locale or region throughout most of their life. For example, oysters are sedentary species.

sedimentation. The process by which suspended, water-borne materials settle out and become deposited on surface substrates. Sedimentation usual occurs when "in-stream" water velocities fall below a minimum speed necessary to maintain particle suspension.

sediment discharge. The deposition of fine particles or sediment within a streambed and into reservoirs. Increased upstream erosion can lead to increased sedimentation leading to shallower river channels and reducing the holding capacity of reservoirs.

seed trees. Trees that are intentionally spared from harvest in order to provide seed stock for forest regeneration on a site.

seemingly unrelated regression (SUR). A method of regression developed by Arnold Zellner where a system of equations is estimated.

SUR accounts for cross-correlation in error terms across equations, and thus provides for more efficient estimators with lower standard errors.

FURTHER READING
Mukherjee, White and Wuyts (1998).

seismic risk. The potential for damage to structures from earthquakes.

selective logging. Timber harvest in which only selected trees are removed from a stand. Trees may be selected based on species, age, form, or other criteria. See *highgrading.*

selfish gene. Term popularized by Richard Dawkins that explains why individuals may co-operate and act in an altruistic way. The selfish gene refers to genetically inheritable altruistic traits in closely related individuals. If such behavior increases reproductive success, then altruism is a "selfish gene" in that altruism conveys no personal advantage to those who display it, but nevertheless the gene for altruism will, over time, become more widespread within a species. See *altruism.*

FURTHER READING
Dawkins (1976).

Sellafield. Site in Cumbria, north-west England, of a nuclear power plant and a nuclear reprocessing facility. The Sellafield facilities have generated a lot of controversy due to the disposal of low-level radioactive waste into the Irish Sea and because of higher than normal radioactivity near the plant. Some believe that the increased radiation is an explanatory factor in the cluster of childhood leukemia in the local area.

semi-renewable resources. Resources that require significant inputs of material and/or energy in order to maintain their initial state such as soil quality, ecological systems, and the assimilative capacity of the environment.

senescence. The dieback of vegetation, often associated with ripening or age.

sensitivity. The extent to which an environment or species responds to changes or shocks.

sensitivity analysis. A method of analysis whereby different results are generated and compared by varying the assumptions and/or parameters used in the model.

separability.   An economic concept used to describe whether individual variables in a function (e.g., utility function or production function) can be aggregated without loss of information about the predicted or dependent variable (e.g., level of utility or output). Separability requires a special set of conditions; for example, a utility function is separable into n groups or aggregates of variables only if the marginal rate of substitution between any pair of variables in any group is independent of the values of variables in any other of the n groups.

septage.   Untreated sewage.

sequence.   A list of numbers from which it is possible to deduce past and future numbers in the series. For example, ... -2, 0, 2, 4 ... is a sequence as it represents a series of numbers where the next number is always 2 greater than the previous number.

sequential games.   A type of game in which players do not move simultaneously, but move in sequence. Often, the choices made by players in previous moves are known to players in later moves.

seral stages.   Distinct plant and animal communities that are in a transitional stage before the climax stage.

seral succession.   The sequential development of a community, especially plants.

sere.   The entire sequence of different plant and animal communities that follow one another in a given area through the process of succession.

serial correlation.   See *autocorrelation*.

series.   A sum of a list of numbers. See *geometric series*.

set.   A grouping of objects called elements.

severance tax.   Tax imposed upon the extraction of resources based upon the volume removed.

Seveso.   A town in northern Italy that suffered from an industrial accident in July 1976 that led to the release of a chemical TCDD that, in turn, contributed to birth deformities and a debilitating skin disease. See *TCDD* and *chloracne*.

sewage. Human and animal excreta that may also contain other contaminants.

sexual selection. The selection of mates with specific characteristics. For example, peahens select peacocks with large and colourful plumage.

shadow price. The price of an asset or resource that reflects its marginal value in use. The shadow price is derived from an optimization problem and will not, in general, equal the market price of the asset or resource.

shallow ecologists. Term coined by "deep ecologists" for persons who may ascribe value to nature based on anthropogenic measures. See *deep ecology*.

Shannon–Wiener diversity index. A widely used index of species diversity within a community. It is defined as the negative of the sum of the relative abundance of each species multiplied by the natural log of its relative abundance. For example, with two species and a relative abundance of 0.5 and 0.5 for each, the diversity index would be $(0.5 \times \log_e 0.5 + 0.5 \times \log_e 0.5) = 0.69$.

FURTHER READING
Stiling (1992).

shared stocks. Renewable resources that are shared between two or more jurisdictions. For example, Canada and the USA share common fishery resources that go back and forth between the two countries' exclusive economic zones.

shellfish. Generic term for mollusks (such as mussels and oysters) and crustaceans (such as lobsters and crabs).

shelter belt. A line of trees or other vegetation planted to protect an area downwind from the prevailing winds.

Shephard's lemma. A mathematical result in which the partial derivative of a cost or expenditure function with respect to input prices can be used to obtain input demands.

shifting cultivation. Also called slash-and-burn agriculture or swidden agriculture. Agricultural practices where land is cleared of trees and grass by burning, or other means. When the fertility of the soil declines to an unacceptable level, the farmers and/or communities move to another area and repeat the practice.

short run.   Economic term that describes the period of time in which only some of the inputs or factors of production can be varied. See *long run.*

sibling species.   Species that strongly resemble each other, but are incapable of breeding.

sic.   Latin term meaning "thus".

sick building syndrome.   Health problems caused by people working or living in buildings with poor air quality.   These may be caused by volatile organic compounds, particulate matter or other pollutants.

side payments.   In game theory, transfers between participants outside of the prescribed payoffs within the game. In co-operative games, such transfers allow participants to share the gains from co-operative behavior.

Silent Spring.   Title of a best-selling book by Rachel Carson which became synonymous with the potential environmental consequences of business-as-usual.

FURTHER READING
Carson (1962).

silt.   Fine soil particles.

Silurian period.   Geological period of time from 438 to 408 million years ago.

silviculture.   The management of trees and forests from planting to pruning to harvesting so as to achieve defined objectives.

simplex method.   A widely used algorithm used to solve linear programming problems.

Simpson diversity index.   An index of biological diversity in an environment defined as

$$D = 1 - \sum p_i^2$$

where p_i is the proportion of individuals from species i.

simulation. Term for calculating the values of variables in a model over time under defined assumptions and parameter values.

simultaneity bias. The bias that exists when a system of equations is estimated, but the simultaneity between variables in the system is not taken into account. Two-stage least squares estimation is an approach that can help overcome the problem.

simultaneous equation model. A set of equations where some of the endogenous variables are determined simultaneously.

sine qua non. Latin term meaning "an indispensable condition".

single-period rotation. See *Fisher rotation*.

site attributes. A list of quantifiable descriptors of the amenities and disamenities of a recreational destination. Site attributes are the primary explanatory variables in models of recreational site choice.

SI units. Système International d' Unités; or more commonly known as the metric system. See Appendix 3.

size limits. Restrictions imposed on harvesters of a resource that prevent them from capturing individuals below a certain size. Size limits are often imposed to protect juveniles from exploitation and to ensure the sustainability of the resource.

skewness. The degree to which a probability distribution is not symmetric. Positive (negative) skewness exists when the distribution has a long tail to the right (left).

SLAPP. An acronym for Strategic Lawsuit Against Public Participation and that is used to describe legal actions filed against activists designed to preempt public protest by proponents of potentially controversial developments and/or actions.

slash. Material left over after the harvesting of trees or removal of vegetation for other purposes.

slash-and-burn agriculture. See *shifting cultivation*.

Slutsky equation. An equation commonly used in economics that shows the change in the demand for a good following a price change in another good in terms of a substitution effect (change in consumption from the price change alone but holding the individual to the same level of utility) and an income effect (change in consumption due to the change in real income because of the price change).

Small is Beautiful. Best-selling book by E.F. Schumacher that argued in favor of small and traditional technologies and practices, and against large-scale industry as a means to promote economic development and welfare.

FURTHER READING
Schumacher (1973).

smog. See *photochemical smog*.

smog trading. See *pollution permits*.

social accounting. A method of accounting that includes allowances for changes in environmental and other socioeconomic variables in calculating national accounts (such as costs incurred from a reduction in environmental quality).

social accounting matrix (SAM). An accounting matrix in which income flows and capital stocks are tracked across all social institutions, including households.

social capital. The institutions and norms of behavior that facilitate social and economic co-operation. Social capital is important in explaining economic performance and environmental quality.

social cost. The private and external costs associated with an economic activity. For example, the social cost associated with the production of a good by a firm that, as a byproducts discharges emissions into the air, would include the private costs of the firm and the costs imposed on the environment and others due to the emissions.

social discount rate. A discount rate that should reflect the social rate of time preference or the social opportunity cost of using public funds. In general, the social discount rate is less than the market rate of interest. See *discount rate*.

social ecology. A school of thought that suggests that many societies' institutions are exploitative and hierachial. According to this view, exploitation on the human dimension is reflected in exploitation of nature, which is the detrimental to the environment.

socially optimal use.   The allocation of resources that will yield the most benefit for society as a whole, taking into account all costs and benefits.

social optimality.   See *socially optimal use.*

social security.   State-supported welfare transfers to poor, disabled and retired members of a society.

social welfare.   1. A measure of the collective well-being of a society 2. Social security.

social welfare function.   A function that supposedly represents the aggregate preferences of individuals in a society. Social welfare functions are used to consider the society payoffs under different sets of policies.

social welfare rule.   A rule, which may be arbitrary, that determines how individual welfare is aggregated across individuals and time so as to find a measure of social welfare.

soft energy.   Term coined by Amory Lovins to describe energy derived from small producers relying on renewable resources, such as solar power, wind power, and hydropower. See *hard energy.*

soft water.   Water that has a low proportion of alkaline compounds and has a low salinity. See *hard water.*

softwoods.   Usually defined as cone-bearing tree species such as pines or firs.

soil.   A mix of organic material, micro-organisms and weathered rocks on which plants grow and obtain nutrients. Soil quality is affected by cultivation and requires careful management to avoid soil erosion and loss of fertility.

soil compaction.   The loss of spaces within the existing soil structure that reduces the productivity of soil and may lead to increased run-off.

soil conservation. Methods and practices designed to avoid or reduce soil erosion and preserve soil fertility. Practices may include terracing on slopes, planting of shelter belts and no-till agriculture.

soil degradation. The diminution in the productive capacity and fertility of the top soil.

soil erosion. The transportation of soil from land due to wind, precipitation and gravity. Erosion occurs naturally but is often increased by improper agricultural practices which unnecessarily expose the topsoil to the physical elements and can reduce soil moisture. Soil erosion is a major and growing problem in many countries.

soil fertility. The inherent productive capacity of the soil.

solar cell. A panel or other device capable of producing an electric current from sunlight. See *photovoltaic electricity*.

solar constant. Average rate at which the earth's surface receives radiant energy from the sun.

solar power/energy. Energy derived from sunlight may be converted into an electrical form through the use of photovoltaic panels, or may be used directly to heat living spaces and or water. See *photovoltaic electricity*.

solar radiation. Radiation from the sun in the form of electromagnetic waves that includes infrared, visible and ultraviolet radiation.

solid load. The material in a water body that is solid material.

solid waste. Waste material, primarily of municipal origin, that is in solid form. The proper disposal of solid waste is an important environmental issue.

Solow growth model. See *neoclassical growth theory*.

Solow–Hartwick rule. See *Hartwick's rule*.

solvent. A fluid that is able to dissolve other substances and bring them into solution.

South, the. A collective term for the developing countries of the Southern Hemisphere.

spaceship earth. Term popularized in the 1960s, originally coined by Kenneth Boulding, that emphasizes that the earth is a closed system in terms of materials (but not energy).

spacing. The selective removal of trees in a stand of an early age in order to promote growth, wood quality, or other forest management objectives.

spatial data. Data that includes observations from at least two different locations and where differences across locations are of interest.

spatial distribution of forest disturbance. The pattern of activities that disrupt tree growth, including fire, disease and harvesting, and which can be characterized by differences in intensity, dispersion and size.

spawning biomass. The total weight of all individuals in a population that is sexually reproductive.

speciation. The process by which new species are formed.

species. The major subdivision of a genus composed of organisms that are genetically and morphologically similar, and are capable of successfully breeding with other organisms of the same species.

species composition. The percentage of a forest stand that is compromised of each species, based upon number of stems, basal area, or gross volume.

species diversity. See *biological diversity*.

species–energy hypothesis. The hypothesis that habitats with greater primary production will support a greater diversity of species.

species extinction. See *extinction*

species hot spots. Area of unusually high species diversity.

species prospecting. See *biodiversity prospecting*.

specific gravity. Density of a substance relative to the density of water.

spectra. The distribution of different wavelengths of extraterrestial solar radiation. The wavelengths associated with various colors become attenuated to varying degrees as they pass through the atmosphere and obstacles such as clouds, water and vegetation.

speculation. Using differences in the prices of goods or assets across time to generate a return; for example, buy at a low price today to sell at a high price sometime in the future.

spill-overs. See *externality*.

spore. A non-active form of a micro-organism that is capable of withstanding extremes in its environment.

sport fisheries. A defined system of recreational fishing encompassing anglers and fish populations harvested for non-commercial, non-subsistence, recreational purposes.

spot market. Market for commodities where delivery from the buyer to seller is immediate.

spot price. Current price in a spot market. The price of crude oil is frequently given as a spot price rather than the price paid by buyers in a long-term supply contract.

spring flush. The rapid melting of snow and ice in the spring in mountain environments and high latitudes. Where the land is contaminated with pollutants and the precipitation is acidic, the spring flush can be very damaging to the environment.

square matrix. Matrix where the number of columns equals the number of rows.

square mile. Unit of area equal to 2.58 square kilometers.

stability. Term used to describe the behavior of a variable that converges or returns to a particular point.

stabilization policy. Government policy designed to reduce fluctuations in national outcome. For example, in a fast-growing economy the government may wish to increase interest rates to reduce the possibility of growth that may be lead to inflation. By contrast, in a slow-growing economy, public works may be initiated to provide income for the unemployed.

stagflation. High inflation coupled with high unemployment.

stand. A group of trees within an area in proximity to one another and relatively homogeneous in terms of age, structure, composition and physical environment. A stand is one of the basic units of forest management.

standard. A rule or regulation set to achieve a desired outcome. In environmental regulations, standards often exist in terms of maximum permissible discharges or concentrations.

standard deviation. A measure of dispersion or spread of values of a variable around its arithmetic mean and defined as the square root of the variance.

standard error. Standard deviation (or square root of the variance) of the distribution of a statistic from a sample.

standing trees. Living, intact trees (in contrast to trees felled by timbering, clearing, fire or natural disturbances).

starting point bias. Term used to describe the potential bias that may exist in surveys designed to elicit individual valuations of an aspect of the environment due to the initial bid supplied in the survey instrument. See *contingent valuation*.

stasis. A period of rest or equilibrium where there is relatively little change in an ecosystem. In high latitudes, winter is a period of stasis relative to summer.

state-determined stock. A stock that is endogenous or determined within a model.

stated preferences. The relative economic desirability of different goods (or combinations of goods) as expressed by an individual.

state planning. Economic plans, sometimes over a five-year period, through which the state attempts to achieve economic and social objectives via administrative fiat rather than markets.

state rights. A property-rights regime where the rights are held by a centralized authority. For example, in many countries state rights exist over national parks.

FURTHER READING
Grafton (2000).

state variable. A variable in an optimal control problem that can only be affected indirectly via control variables and which represents the "state of nature". For example, in an optimal control problem of a fishery the size of the fish population is the state variable while the control variable would be the level of harvest. See *control variable* and *optimal control theory*.

static efficiency. See *economic efficiency*.

static reserve life. The number of years before an economically extractable (mineral) reserve is exhausted at current rates of extraction.

FURTHER READING
Tietenberg (1996).

static sink life. A measure of resource extraction that accounts for the fact that metals extracted from the earth's crust end up on its surface.

FURTHER READING
Holmberg, Robert and Erikson (1996).

stationarity. A variable the mean and variance of which are time invariant. Most time-series economic data are non-stationary such that the expected value of the variable depends on time. In regression analysis, using time-series data where the variables are non-stationary, spurious correlations among variables may arise. To correct for this potential problem, procedures have been developed to derive stationary variables from non- stationary variables. See *differencing*.

stationary point. Also known as a critical point in mathematics, this is where a function is neither increasing nor decreasing such that the first derivative is zero.

stationary resource. A natural resource that is not mobile, such as a forest.

stationary source. A source of pollution that is fixed in location, such as a factory. See *mobile source*.

stationary state. A point where the value of a function is unchanging with respect to time.

statistical efficiency. Term used to compare statistical models where the estimate of a parameter with the lowest variance is the most efficient. See *variance.*

statistical life. A unit of statistical measure corresponding to the average lifespan for an average individual. Statistical lives are used in aggregate measures such as income/statistical life or value/statistical life.

status quo. Latin term for the current state of affairs.

Steady state. A point at which a dynamic system is in a state of equilibrium.

steady-state economy. Term popularized by H.E. Daly (1973) that refers to an economy that experiences no change in its stocks of capital and wealth, no change in its population, and a minimum flow of throughputs so as to minimize environmental impacts.

stem volume. The volume of the principal or main woody stem of a plant.

steppe. A biome of grasslands characterized by the feature of low rainfall and extreme ranges in temperature.

stochastic models. Models the outcome of which are determined, in part, by stochastic processes. See *stochastic process.*

stochastic process. Term used to describe the process where random variables may take on a range of values, called the state space, over a given period of time.

stock. The quantity of a resource at a given point of time. For example, the biomass of a fishery is a stock, but the harvest of fish is a flow. See *flow.*

stocking rate. Rate at which a given area of land is used in terms of the number of animals grazed, or the number of trees growing per hectare.

stock pollutant. A pollutant, such as a PCB, that accumulates in the environment. See *polychlorinated biphenyls.*

stock-recruitment model. Model that relates the stock or biomass of a population, or the spawning biomass, with the number of recruits. Stock-recruitment models are used in fisheries and they usually suppose that the smaller the stock, the greater is the rate of recruitment.

FURTHER READING
Hilborn and Walters (1992).

storable resource. A natural resource that can be stored or accumulated for future use, such as water.

straddling stocks. Resources that are found in the jurisdiction of one state but can also be exploited outside of its jurisdiction. For example, most of Canada's fishery resources lie within its 200-nautical-mile exclusive economic zone, but some of its fisheries (nose and tail of the Grand Banks) lie outside its direct jurisdiction, or that of any other country.

strange attractor. A set of points to which a chaotic dynamic system converges and which are highly sensitive to the initial conditions. See *chaos.*

strategic bias. In contingent valuation, the potential that a respondent's stated willingness to pay (or accept) may be different from her true willingness to pay (or accept) if she believes that her answer will influence the provision of the good or service in question. See *contingent valuation.*

strategic form. Also called a normal form. A graphic description of any game involving two different participants using a table made up of different cells where one participant's strategies are displayed in horizontal rows and the other in vertical rows of the table. The resulting payoffs to each participant from a combination of the different strategies are then found within the cells formed by the intersection of the two strategies.

strategic interaction. The strategies or behavior pursued by a participant that benefits others, but that also serves to improve the participant's own well-being.

strategic petroleum reserve (SPR). Reserves of petroleum of about 600 million barrels of crude oil owned by the US government and stored in case of a major disruption to oil supplies. See *buffer stocks.*

strategic plan. A land allocation plan that provides both land-use objectives and implementation strategies.

strategic rationality. Choosing a strategy that is consistent and believable to the other parties.

stratified random sampling. Random sampling of observations in defined strata or classifications. The approach is widely used to provide adequate representation of strata (such as income levels) of interest.

stratosphere. A region in the earth's atmosphere from 10 to 50 kilometers above the earth's surface (as low as 8 kilometers in polar regions and as high as 15 kilometers in tropical regions) and which contains the ozone layer that helps shield the earth's surface from ultraviolet radiation. See *stratospheric ozone.*

stratospheric ozone. Ozone found in the stratosphere and which protects life on the earth's surface from ultraviolet radiation from the sun. See *ozone hole.*

stream flow. The amount of water passing through a stream channel and typically expressed in volumetric units over a specific period of time.

stream load. The amount of material transported by a stream or river in a given period of time.

strict liability. A legal term that holds individuals or firms liable for their actions, even if their actions were not negligent. See *liability.*

strip-cropping Growing crops in raised strips that follow the contour of the land so as to reduce erosion.

strip cut. Pattern of harvesting wood in long narrow strips.

strip mining. Open-pit mining characterized by the removal of the overlying soil and rock.

strong sustainability. The principle that there should be no net depletion of natural resources from economic activity. This contrasts with weak sustainability, which allows for the conversion of one type of capital stock into another type, so long as the aggregate value of the net capital stock does not decrease.

FURTHER READING
Neumayer (1999).

structural diversity.   Variations in the physical landscape (such as a hill or a lake) that increase biological diversity.

structural equations.   Equations in a simultaneous equation model that include equilibrium conditions (such as supply equals demand) and behavioral equations (such as a supply equation and a demand equation).

student's t-test.   See *t-test.*

stumpage price.   See *stumpage value.*

stumpage rate.   The charge that is paid to the owner for the right to harvest a given volume of wood from a forest.

stumpage value.   The value placed upon a tree on the stump. Generally, stumpage is the market value of a log less transportation, milling, and harvesting costs.   See also *market stumpage prices.*

subalpine.   The habitat zone found just below the alpine zone (differentiated by the limit of tree growth).

subcanopy.   Layers of forest vegetation occurring beneath the forest canopy.

subeconomic resource.   A natural resource for which the marginal cost of extraction or harvest exceeds the marginal benefit of extraction.

subgame perfect.   A concept in game theory in which a particular strategy holds not only for the equilibrium outcome, but also for any of the other possible outcomes.

FURTHER READING
Binmore (1992).

sub judice.   Latin term for an issue or case before the courts.

sublittoral zone.   The environment found below the littoral zone and marked by a transition from aquatic plants to blue-green algae below which light does not effectively penetrate. See *littoral zone.*

sub-Saharan Africa. African countries located south of the Sahara desert.

subsidence.   Fall or depression in the earth's surface. Too rapid a rate of groundwater withdrawal can be a major cause of local subsidence.

subsidy.   Any payment from a central authority to producers or firms that reduces their costs of production. Subsidies frequently exist in agriculture in developed countries where many governments provide price and income supports to farmers.

subsistence.   The ability to survive or subsist but with no economic surplus or savings should circumstance worsen. Persons living at the subsistence level have few degrees of freedom in terms of their decisions and do not have the resources to ensure their own survival or the sustainability of the environment and the resources from which they obtain their livelihood.

subsistence agriculture.   See *subsistenc.*

sub-species.   Different groups of organisms with different physical characteristics or behavior, but which are capable of interbreeding.

substitute.   A good or commodity that is consumed or produced in place of another good. For example, some people drink tea as a substitute for coffee.

substitution effect.   The change in consumption of a good following a price change due to the substitution to other goods, but holding the consumer's utility unchanged. See *income effect.*

subtidal kelp.   Kelp found below the intertidal zone.

subtractable.   See *rivalrous.*

suburbanization.   The conversion and incorporation of small communities and towns into suburbs of larger urban areas. See *urbanization.*

succession.   The orderly and predictable development of distinct stages within a given area, and marked by different species and processes over time.

succulent.   A plant with thick leaves and stems suited for the retention of moisture.

sufficient conditions.   Conditions that, if satisfied, ensure a particular outcome. Sufficient conditions are particularly useful in the solution of optimization problems. See *necessary conditions.*

suffrutex.   A shrub often found in the understory of (African) forests.

sulfur.   A non-metallic element and commonly occurring solid element, the compounds of which are one of the principal constituents of acid rain and atmospheric pollution, but it is also a necessary part of many organic processes.

sulfur dioxide (SO_2).   A gas formed when sulfur is burnt in the presence of oxygen, and a major anthropogenic cause of acid rain. An important source of sulfur dioxide is coal-fired electric utilities that burn sulfurous coal to produce electricity.

sulfur dioxide trading.   A system of pollution permits, or tradable emission permits, established in the USA to control the emissions of sulfur dioxide from coal-fired electric utilities. The total allowable level of emissions has been declining so as to try to achieve environmental objectives and to control the problem of acid rain in eastern North America. Under the system, firms can meet their individual emission obligations by installing pollution control devices and/or purchasing emission permits from other firms.   Trading of emission permits allows high-cost-of-abatement firms to reduce their costs in meeting their emission targets. See *pollution permits.*

sunk cost.   A forgone expenditure incurred in the past but which has no effect on present or future returns.

sunspots.   Spots on the sun that are associated with strong electro-magnetic activity. Changes in sunspot activity, which occur in regular cycles (of approximately 11 years), have been linked to changes in weather and solar radiation.

superfund.   See *CERCLA*.

superfund sites.   US hazardous waste sites defined under the Comprehensive Environmental Response, Compensation and Liability Act (CERCLA).

superphosphate.   A chemical fertilizer prepared with either a combination of sulfuric acid, calcium sulphate, and calcium acid phosphate or phosphoric acid.

supply.   The quantity available for sale of a good or service at a given point in time and at the market price.

supply curve.   See *supply function*.

supply function.   A relationship between the amount firms are willing to supply, or produce and market of a good or service, and the price of the good.

supra.   Latin term meaning "above".

supralittoral zone.   The area above the littoral zone that is influenced by the spray from waves.   See *littoral zone*.

surface water.   Water found on the earth's surface in streams, lakes and other surface features.

surplus forest.   Forest where the annual cut is less than the long-term yield from a forest.

surplus production model.   A model of population dynamics in which the next period's biomass is calculated as the current period's biomass plus surplus production (recruitment and growth less natural mortality) less the amount harvested.

FURTHER READING
Hilborn and Walters (1992).

survival of the fittest.   Term coined and popularized by Herbert Spencer, a British scientist of the nineteenth century. The term is often interpreted to mean that Darwinian evolution has led to higher order and superior life forms, the ultimate example being *Homo sapiens*. In reality, natural selection simply favors individuals (and their offspring) that are adapted to their particular environment and does not necessarily lead to larger or more intelligent species. See *natural selection*.

FURTHER READING
Gould (1993).

suspended solids (SS).   A mixture of solids carried within a fluid.

sustainability.   In general, the idea that the natural capital stock remains constant over time. Differences in what constitutes the natural capital stock and what exactly remains constant over time have led to a number of different definitions of sustainability. See *weak sustainability* and *strong sustainability*.

FURTHER READING
Gale and Cordray (1994).

sustainability criterion.   The principle that our actions should not make future generations any worse off than the current generation.

sustainability paradigm.   A program of resource use over time that modifies traditional cost–benefit approaches to maintain either constant capital assets (weak sustainability paradigm) or constant critical natural capital assets (strong sustainability paradigm).

FURTHER READING
Neumayer (1999).

sustainable.   The notion that an aspect of the environment or practice can be maintained into the foreseeable future.

sustainable agriculture.   Methods of farming that do not degrade the productive capacity of the land.

sustainable development.   A term popularized by the United Nations in the report *Our Common Future,* and which is defined in many different ways. It is commonly defined as a process by which current generations meet their needs and care for the environment in ways that do not compromise the needs of future generations.

FURTHER READING
Rao (2000).

sustainable forest management (SFM). Forestry management that allows for the harvesting of timber, but also ensures that the forest provides needed ecological and landscape functions.

sustainable tourism. Tourism that is designed to have minimal environmental impacts.

sustained yield. Yield or harvest from a renewable resource that can be sustained into the future without reducing the overall size of the resource.

sweatshops. Factories or plants where the workers are employed in sub-standard conditions.

swidden. An agricultural plot cleared by slash-and-burn methods.

swidden agriculture. See *shifting cultivation*.

symbiosis. A mutually beneficial relationship between organisms. See *mutualism.*

sympatric speciation. The evolution of new species, without geographical separation, due to inbreeding or reproductive isolation of a segment of the population.

synergistic effects. When two or more substances work together to produce a greater effect than would be expected from simply adding the individual effects from each of them together.

synfuels. Shortened form of synthetic fuels and defined as liquid petroleum products that are synthesized from hydrocarbons, such as coal or tar sands.

synthetic chemicals. Chemicals synthesized artificially in laboratories or factories rather than by living organisms.

system. The interconnections between organisms and physical processes that together represent an integrated picture of the various sub-components.

systematics. The classification of flora and fauna.

system dynamics. An approach to understanding dynamic phenomena that involves modeling of the stocks, flows and feedbacks of a system and the simulation of the models under various scenarios.

system of national accounts (SNA). See *national income and product accounts*.

T

taiga. See *boreal forest*.

tailings. Crushed waste material and rock left from mining. Without some form of remediation tailings can often have an adverse effect on the environment.

takings. 1. Under the Endangered Species Act in the USA, a taking is defined as the killing, harming, or harassment of an endangered species. 2. A governmental appropriation of property rights, either in part or complete, such as the restrictions placed upon the allowable uses of land at a particular site. See *incidental taking*.

tall stacks. Large stacks up to several hundreds of meters high that are used to disperse air pollution and to minimize its impact on the local area. Tall stacks often result in greater spatial dispersion of the pollution.

tannins. Toxic substance found in plants that renders them bitter or astringent.

targeted species. A preferred species that is actively sought in the process of harvesting or capture. See *bycatch*.

tariff. A tax imposed on imports that is often based as a percentage of the import price.

tar sands. Name given to sand deposits that contain heavy oil. Tar sands represent a huge reserve of hydrocarbons equivalent to 400 billion barrels of oil.

tax. See *Pigouvian tax*.

tax incidence. Term for the distribution of the burden of a tax such that those who initially pay the tax (such as a wholesaler with a wholesale tax) are not necessarily the persons who end up paying for the tax (consumers) after price adjustments.

taxonomy. See *systematics*.

Taylorism. Term describing production processes in which a distinction is drawn between management and labor, and where managers establish regimented production practices to be followed by laborers, often using assembly-line techniques. Typically associated with early mass-production processes. See *Fordism*.

TCDD. Acronym for 2,3,7,8-tetrachlorodibenzo-p-dioxin that is a toxic substance used in the manufacture of herbicides. See *Seveso*.

technical change. See *technological change*.

technical efficiency. The ability of an enterprise to produce the maximum amount of output from a given amount of inputs. Enterprises that are technically inefficient have the potential to increase their output without increasing their inputs. See *allocative efficiency, economic efficiency*, and *scale efficiency*.

technical progress. See *technological change*.

technological change. Improvements in production processes over time which permit more outputs or services, or an improved quality of goods and services, for a given level of inputs. See *disembodied technological change* and *embodied technological change*.

technological externality. An externality that distorts the efficient allocation of inputs and/or outputs. See *externality* and *pecuniary externality*.

technology. The aggregate knowledge of production processes that can be currently applied to produce goods and services.

technology-based standard. A command and control regulatory approach in which the method of pollution abatement is legally mandated and monitored by the government. See *command and control*.

technology forcing. Standards that require the adoption of a particular type of technology.

technology transfer. The transfer of technological processes or knowledge from one market or nation to another.

temperate rainforest. Forests found in temperate regions of the world and distinguished from other temperate forests by the infrequency of fire, the predominance of evergreen species, and a high degree of structural complexity.

temperature dynamics.   Changes over time in the temperature of the earth's surface and atmosphere due to perturbations in the earth's radiation budget.

temperature inversion.   See *inversion.*

tender.   An offer to contract or supply a good or service at given level of quality, in a specified time period, and at a specific price.

tenure.   A broad term that represents the property rights over land and, in particular, the duration of these rights.

TEQ.   A method of measuring the toxicity of a combination of furans and dioxins and other substances relative to the toxicity of TCDD.   See *TCDD.*

teratogen.   Any substance that leads to physical and mental deformities in a growing fetus.

terminal time.   The last period in a dynamic optimization problem.

terminal value.   The value of a project or asset at terminal time.

terms of trade.   Ratio of export prices to import prices. A long-term decline in the terms of trade reduces the ability of countries to pay for their imports.

terracing.   The creation of level planting areas that follow the contours of elevated land and are designed to reduce soil erosion.

terrestial radiation.   The long-wave infrared radiation emitted by the earth.

Territorial User Rights in Fisheries (TURFs).   Term   coined   by Francis Christy to denote areas of the ocean or sea floor to which individuals or groups of individuals have property rights in terms of the harvest of marine animals. TURFs are not widely used in fisheries although they do exist for some shellfish. See *individual transferable quotas.*

Tertiary. A geological time period which began 66.4 million years ago and lasted until about 1.6 million years before the present.

tertiary sewage treatment. A treatment beyond primary and secondary treatment that may involve a number of processes including reverse osmosis, carbon filters and other filters to remove nitrates and chemicals from treated wastes. See *primary sewage treatment* and *secondary sewage treatment*.

tetraethyl lead (TEL). A gasoline additive used in leaded gasoline so as to improve engine performance. Concern over lead pollution has led to a phase-out of the use of TEL in many countries.

thalidomide. Prescribed drug widely used in the 1960s that resulted in severe physical abnormalities in infants exposed to the drug as fetuses.

theory of the second best. Theory which states that if not all the conditions for economic efficiency are attained in an economy, it may not be opti-mal or desirable to ensure economic efficiency in all other conditions or sectors.

FURTHER READING
Lipsey and Lancaster (1956).

thermal expansion. An increase in water volume due to an increase in its temperature. Climate change that leads to increased surface temperatures will reduce the density and increase the volume of the water in the world's oceans. This effect is more pronounced at higher water temperatures and could result in a rise in sea level of about 30 centimeters by 2100 if average surface temperatures increase by 2.5 degrees Celsius over the century (Houghton, 1997).

FURTHER READING
Houghton (1997).

thermal pollution. An increase in temperature caused by anthropogenic or natural source that can have a negative effect on an ecosystem. For example, the discharge of hot water into streams can be responsible for fish kills.

thermocline. A region marks the boundary between layers of water which have different temperatures and densities. Thermoclines exist in ponds and lakes, and also oceans.

thermodynamics.   The study of different forms of energy, and especially the conversion of heat into other forms of energy, and vice versa. See *law of conservation of energy* and *second law of thermodynamics*.

thermophile.   Bacteria that live in extreme environments in terms of heat, able to exist in temperatures of up to 80 degrees Celsius. See *hyperthermophiles*.

thinning.   The removal of trees from an immature crop so as to promote growth and form among the remaining trees.

thionic flucisols.   Alluvial, estaurine soils containing large concentrations of sub-soil sulfuric acid when dried.

Third World. Collective term coined for developing countries.

33/50 program.   A USA government voluntary compliance program to reduce emissions of specified pollutants by firms by a third by 1992, and by 50 percent by 1995.

FURTHER READING
Callan and Thomas (2000).

THMs.   See *trihalomethanes*.

threatened species.   Species that may become endangered should current trends continue, and may be in danger of extirpation in some environments.

Three Gorges project.   A massive hydro-electric development on the Yangtze river in China that will displace at least 1.5 million people and will flood a unique area of beauty and historical significance including the Xiling, Wu and Qutang gorges.

Three Mile Island.   Location of a nuclear power plant in Pennsylvania, which suffered a nuclear accident and resulted in the temporary evacuation of over a hundred thousand people from the local area.

threshold effect.   An effect which does not arise until after a given dose or level of pollution. Many pollutants have threshold effects which will vary depending on the environment and the individuals exposed to the pollutants.

threshold limit values (TLVs).   A standard level of exposure to which most people can be exposed over time without perceived health risks. TLVs are used to help develop safe working practices for workers.

tidal power.   The generation of power with electrical turbines driven by changes in sea level due to the influence of tides.

Tiebout hypothesis.   Named after its formulator, this suggests that communities and jurisdictions can compete for residents or workers by offering them a desirable mix of public services and public goods. Jurisdictions that fail to provide a desirable mix of tax and public goods services may lose out in terms of migration of their residents to more desirable jurisdictions.

FURTHER READING
Cornes and Sandler (1996).

till.   Sediment deposited by glaciers.

timber concessions.   Rights to harvest public timber within an area granted by the government.

timber line.   The altitude or latitude beyond which trees cannot grow because of the severity of the climate.

timber utilization.   The quantity and quality of timber harvested from a defined area of land. A low utilization rate often implies less sustainable forest management practices, as it often coincides with greater on-site wastes.

time preference.   1. A preference for consumption in the present rather than in the future. 2. The relative weight given to consumption, utility, or welfare in any one period relative to any other period.

time-series data.   Data where observations of variables are available over time or multiple time periods.

tipping fees.   Charges paid to waste disposal sites or landfills for the right to dispose of specified wastes.

Tobit model.   A method of estimation due to James Tobin that estimates a regression equation where the data is "censored" or cut off at a particular value.

FURTHER READING
Ramanathan (1992).

tolerance.   Ability of an individual, species or ecosystem to maintain itself under a given set of conditions. For example, shade-tolerant species can survive in conditions of shade.

topsoil.   The upper and most fertile level of soils, which contains most of the organic material.

Torrey Canyon.   Name of oil tanker than ran aground off the coast of England in 1967 spilling over 100,000 tons of crude oil, and killing many thousands of marine birds and fish.

total allowable catch (TAC).   Terms used in fisheries to denote the total permissible catch of the fishing fleet in a given period of time, usually a fishing season. The TAC is used to ensure the sustainability of the resource.

total cost.   All costs incurred in producing a good, including both variable and fixed costs. See *total variable cost* and *total fixed costs*.

total dissolved solids (TDS).   A measure of the total quantity of solids dissolved in a solution, this is an indicator of the level of water pollution.

total economic value.   The total willingness to pay to preserve, maintain, or possess a natural resource or environmental good. The total economic value includes use and non-use values.

total factor productivity (TFP).   A broad measure of productivity defined as a ratio of output(s) to all inputs. See *productivity*.

total fixed costs.   The total costs of a firm that do not vary with the level of output.

total particulate matter.   A measure of the total particulates in a given volume of air, used as an indicator of the level of air pollution.

total suspended solids (TSS).   A measure of the solids suspended in a liquid, commonly used as an indicator of water pollution.

total variable cost.   The total costs of a firm that vary with the level of output.

total willingness to pay.   The maximum transfer, in income or wealth, that an individual would make in order to ensure a certain state of the world (e.g., the possession of a good, access to a benefit stream, the continued existence of a resource, etc.).

toxicity.   The degree to which a substance can be poisonous.   It includes acute toxicity (the effects due to a single exposure) and chronic toxicity (the effects due to several small-level exposures).

toxic pollutants.   Materials that, upon exposure, have a detrimental effect on an organism's health.

toxic release inventory (TRI).   A US database compiled out of a requirement that large manufacturers report their annual releases of defined toxic chemicals.

toxigenic E. coli.   See *Escherichia coli O157:H7.*

toxins.   Organically produced poisons that are usually a byproduct of bacteria.

trace gas.   A gas found in trace amounts, in terms of greenhouse gas concentration, in the atmosphere. For example, the most important, carbon dioxide, currently represents about around 0.035 percent of the atmosphere.

tradable discharge permits (TDPs).   See *pollution permits.*

tradable effluent permit market.   The real or theoretical market in which emissions permits or quotas can be traded freely in order to achieve a least-cost provision of a fixed level of ambient environmental quality. See *emission permit.*

FURTHER READING
Tietenberg (1996).

tradable emission permits (TEPs).   See *pollution permits.*

tradable permits. See *pollution permits*.

tradable SO$_2$ emission allowances. The tradable right to produce a specified quantity of airborne sulfur dioxide. See *pollution permits* and *sulfur dioxide trading*.

FURTHER READING
Tietenberg (1996).

tradeoff. A conscious decision to forgo the benefits of one resource, good, or service, in order to enjoy those of another exclusive resource, good, or service. Tradeoffs also can be made between management objectives.

Trade-Related Intellectual Property Rights Agreement (TRIPS). A World Trade Organization (WTO) agreement that provides the minimum standards of protection of intellectual property that must be undertaken by all members, such as providing copyrights for books or patents for products. A critique of TRIPS is that it fails to provide adequate protection for traditional and indigenous knowledge.

trade winds. Prevailing winds that blow from the north-east in the northern hemisphere and the south-east in the southern hemisphere towards the equator from the mid-latitudes, and which, historically, were extensively used by sailing ships.

traditional knowledge. The body of knowledge regarding the functioning of the physical world drawn from observation and traditions passed on within different cultures. Often characterized as being distinct from information drawn from scientific inquiry.

tragedy of the commons. See *commons, tragedy of*.

transactions costs. The costs incurred in exchanging and enforcing property rights. High transactions costs can be a major impediment to trade and exchange. See *Coase Theorem*.

transboundary pollutants. Pollutants, such as sulfur dioxide, that are produced in one country but that impose environmental harm on other nations.

transects. Linear paths across the landscape used in obtaining data or samples for analysis.

transect surveys.   Data or samples taken along a straight line following a particular compass bearing.

transferability.   The ability of the holder of a property right to sell, exchange or bequest that property to another person or persons.

transfer earnings.   The payment required to keep a factor of production in its current activity or employment.

transfer payments.   Payments made by governments to individuals because of needs such as unemployment benefits, disability payments, and old-age security.

transfer pricing.   The pricing of resources or materials by vertically integrated companies to minimize tax liabilities.

transformation curve.   See *production possibilities frontier*.

transformative values.   The properties of an environmental asset that cause preferences of an individual for that asset to change upon exposure to the asset.

FURTHER READING
Brennan (1992).

transgenic crops.   Plants produced by genetic engineering and that have genes from more than one species. See *genetically modified*.

transition fuels.   Non-renewable resources other than oil and gas, such as coal and uranium, that can provide an interim source of energy supply until renewable energy becomes more widely used.

transitivity.   A requirement for logical preferences where if a bundle of goods A is preferred to bundle B, and bundle B is preferred to bundle C, it cannot be that bundle C is preferred to bundle A.

transpiration.   An important part of the water or hydrological cycle by which plants take up moisture from the soil through their root system and discharge water vapor through their leaves.

transversality conditions.   Terminal time conditions that must be satisfied at the end of a dynamic optimization problem. In the dynamic optimization problem to maximize the present value of rents from a mine, a transversality condition is that at the completion of the program (terminal time) the product of the remaining reserves and the shadow price of the resource must be equal to zero. In other words, either all the resource is extracted or it is no longer worthwhile to extract the resource.

travel-cost method.   The econometric estimation of a travel cost model in which a demand function is derived in order to estimate the welfare contribution of visits to a recreational site, or attributes of a recreational site.   See *travel-cost model.*

travel-cost model.   A model of recreation demand in which the good supplied (or acquired) is the number of trips to a recreation site and the cost of acquisition is travel cost (including out-of-pocket costs and the cost of travel time).

trawl.   See *trawler.*

trawler.   A fishing vessel that employs a net (or series of nets) that is towed behind the vessel along the sea bottom or in mid-water. Trawlers commonly target groundfish species, such as cod and pollock.

treaty rights.   Rights accorded to First Nations or aboriginal people, often in return for access to land or the ceding of resources and land. Treaty rights may include access rights (the ability to continue traditional activities on ceded lands) or direct rights to land and/or resources.

tree ring records.   The use of core samples from long-lived trees to establish climatic records for the past.

Triassic.   Geological time period from 245 to 208 million years ago.

trihalomethanes (THMs).   A possible carcinogen found in some drinking water that is a byproduct of chlorination of water that has high levels of organic matter. In the USA, the maximum contaminant level of THMs is set at 0.1 milligram per liter.

trip limits.   Limits imposed on the amount of a resource that can be harvested per trip. Trip limits are imposed in some fisheries in an attempt to discourage fishers from investing in larger vessels and increasing the capacity of the fleet.

TRIPS.   See *Trade-Related Intellectual Property Rights.*

trolling.   A method of fishing whereby lures are attached to lines that are attached to a vessel and is dragged through the water. Trolling is used to catch salmonids, and also for pelagic species.

trophic level.   The level or stage in the food chain at which an organism feeds.   For example, carnivores feed at a higher trophic level than do herbivores.

trophic subsidy.   The difference between the total amount of energy assimilated by green plants, and the amount of energy used by plants which is converted into organic material.

tropical.   Pertaining to a location on earth falling between 23.5 degrees North (tropic of Cancer) and 23.5 degrees South (tropic of Capricorn) latitude.

tropical dry forest.   Usually deciduous forests in tropical latitudes characterized by a marked dry season.

tropical hard-woods.   Deciduous tree species found in the tropics that are often commercially harvested.

tropical rain forest.   A dense evergreen forest in tropical latitudes receiving at least 2.5 meters of rain annually.   Tropical rain forests are noted for their high biological diversity.

tropics.   See *tropical.*

tropism.   The direction of growth of a plant in response to an external stimulus, such as light or gravity.

tropopause.   A layer in the earth's atmosphere ranging between 10 to 15 kilometers above sea level that represents a boundary layer between the troposphere and stratosphere.

troposphere.   Part of the atmosphere found at the earth's surface and up to 15 kilometers (8 kilometers in polar regions) above the surface. Unlike other parts of the atmosphere, temperature declines as the distance from the earth's surface increases.

tropospheric ozone.   Ozone located at or near the earth's surface, and that can be harmful to plants and animals. See *ozone* and *volatile organic compounds.*

trypanosomiasis. Commonly known as sleeping sickness, this is a disease endemic to many parts of Africa. The disease is contracted through the bite of the tsetse fly and is caused by a protozoa. The disease leads to various symptoms, and is frequently fatal.

t-test. A statistical test used to test the null hypothesis that a coefficient in a regression analysis is significantly different from zero. The degrees of freedom in the test equal n – K, where n is the number of observations and K is the number of regressors.

FURTHER READING
Greene (1997).

tuna-dolphin dispute. A case that involved a US law that prohibited the importation of tuna from countries where the incidental killing of dolphins exceeds a critical amount. In 1988, a lawsuit by a US environmental group argued that the USA was not upholding the law in regards to tuna imported from Mexico, where Mexican fishers used purse seine nets to catch tuna. A US federal court agreed and prohibited any further imports. The Mexican government argued that its right to sell tuna in the USA had been violated and filed an appeal under GATT. The panel that heard the appeal in 1991 agreed with Mexico, but the USA and the Mexican government reached an agreement before the GATT ruling was formally adopted.

tundra. A biome found in extreme northern latitudes north of the boreal forest. Most of the vegetation consists of grass, lichen and moss, beneath which is permafrost. See *permafrost.*

turbidity. The cloudiness in water that is usually a function of the level of suspended solids.

TURFs. See *Territorial User Rights in Fisheries.*

turnpike theorems. Theorems employed in optimal growth theory that suggest that the optimal path arch towards the terminal point rather go there directly.

FURTHER READING
Takayama (1985).

2,4-D. 2,4-dichlorophenoxyacetic acid, a widely used broad-leaf herbicide.

2, 4, 5-T. A herbicide for controlling weeds. Its use is restricted in several countries because of concerns over the release of toxic dioxins that arise from the production and use of the herbicide.

two-part price. Pricing structure where the per unit price paid by consumers varies according to the amount consumed. For example, consumers may initially pay high price per unit, but at higher levels of consumption the price per unit may fall.

two-stage least squares (2SLS). A method of estimating simultaneous equation model, developed by H. Thiel. In the first stage, a reduced form of the model is estimated using ordinary least squares, which saves the predicted values of the endogenous variables. In the second stage, the structural equations of the model are estimated but the observed endogenous variables are replaced by their predicted values.

FURTHER READING
Ramanathan (1992).

two-tailed test. Statistical test that uses both critical regions when testing a null hypothesis. For example, $H_0 : b_1 = 0$ is a two-tailed test if the alternative hypothesis is either $b_1 > 0$ or $b_1 < 0$. See *one-tailed test.*

Type I error. Sometimes called the alpha risk, or significance level, of a statistical test. Type I error is the probability of rejecting a null hypothesis which is true. All things being equal, the smaller the type I error, the greater will be the type II error in a statistical test. See *Type II error.*

Type II error. Sometimes called the beta risk. The type II error is the probability of failing to reject a null hypothesis which is false. See *Type I error.*

typhoid. Bacterial disease that results in high fever and sometimes death. Typhoid is endemic many developing countries and is contracted by eating contaminated food or drinking contaminated water.

U

ultraviolet (UV) radiation. Electromagnetic energy which has a wavelength just beyond the visible light spectrum. Much of the sun's UV radiation is blocked in the earth's atmosphere by stratospheric ozone, but a proportion still reaches the earth's surface. Long-term exposure to UV radiation can reduce immune responses in humans and cause skin cancers and other ailments. Ultraviolet A, B and C radiation have wavelengths of 320–400, 280–320 and 200–280 nanometers.

uncertainty. A situation where many possible events or outcomes may occur, but it is not possible to assign probabilities to these events in advance.

underemployment. Persons who have some form of employment, part-time or seasonal, but who are willing and able to work more.

understory. The layer of vegetation made up of shrubs and plants on the forest floor.

undervaluation. Placing too low an economic value upon an activity or resource without taking into account all the economic benefits associated with that activity or resource. For example, tropical forests yield not only timber products that tend to be underpriced for institutional reasons, but also yield a variety of non-timber products and other environmental values (such as biodiversity and carbon sequestration) that are not captured in timber prices.

unemployment rate. A measure of the number of people out of work but willing to work and looking for work. The unemployment rate is measured as a proportion of the total workforce.

UNESCO. See *United Nations Educational, Scientific and Cultural Organization.*

uneven-aged stand. A forest in which at least three stands are of significantly different ages (a difference of greater than 20 years).

ungulate.   A group of herbivore mammals with hooves, such as horses and deer.

uniform distribution.   A probability distribution or density function where the probability of picking any value of the random variable, defined by the distribution, is identical.

uniformitarianism.   The notion that the current physical processes also existed in the geological past, and that observation of current events can provide insights about the development of the Earth.

unilateral co-operation.   The strategy followed by a participant in a game who chooses to co-operate with others without any assurance that such co-operation will be reciprocated.

unilateral defection.   The strategy followed by a participant in a game who chooses to deviate from any co-operative arrangement and pursue his or her own self-interest.

unimodal curves.   Functional relationships that exhibit an "inverted U shape" (with a single global maximum), or a U shape (with a single global minimum).

unit cost measures of scarcity.   Measures of the scarcity of non-renewable resources based on the costs of production or extraction.

United Nations Conference on Environment and Development (UNCED).   The 1992 conference that took place in Rio de Janeiro, Brazil that resulted in an environmental action plan called Agenda 21 and the United Nations Framework Convention on Climate Change, the Convention on Biological Diversity and the Convention to Combat Desertification.

United Nations Convention on the Law of the Sea (UNCLOS).   A convention developed by a series of international meetings in the 1970s and 1980s to govern the exploitation of marine resources. The Convention established 200-nautical-mile exclusive economic zones (EEZs) from the coastline of maritime states.

United Nations Development Program (UNDP).   A United Nations (UN) agency charged with assisting other UN agencies in undertaking social and economic development in poor countries.

United Nations Educational, Scientific and Cultural Organization (UNESCO).   Formed in 1946 and charged with fostering international co-operation of scientific research. Among its many activities, UNESCO directs the Man and the Biosphere program.

United Nations Environment Program (UNEP).   Formed in 1972, the UNEP is based in Kenya and co-ordinates international efforts at monitoring and protecting the environment.

unitization.   An agreement over the use of common-pool resources, and particularly over oil fields, in which the owners of the sub-surface rights agree to let one user extract the resource and provide net benefits to all others at an agreed-to formula.

FURTHER READING
Libecap (1989).

unit root tests.   A statistical method used to test whether variables are stationary by evaluating whether the variable is following a random walk. If the null hypothesis that the variable is following a random walk cannot be rejected, then the variable is non-stationary.

FURTHER READING
Mukherjee, White and Wuyts (1998).

universal property rights.   Situation where all property rights have been completely assigned.

unstable equilibrium.   A condition in dynamic analysis when a system will not return to its original state following a perturbation. In biological systems, the minimum viable population stock is an unstable equilibrium because a positive perturbation will cause stocks to grow to the carrying capacity, while a negative perturbation leads to extinction.

upwelling.   A place where cold and often nutrient-rich water approaches the sea surface, characterized by increased marine productivity.

uranium.   A metal commonly found as two isotopes, uranium 238 and uranium 235. Uranium 235 is used as a fuel in nuclear fission. See *nuclear fission*.

urban heat effect or island.   Phenomenon where the heat sources, building materials and possibly air pollution of a city can help keep its temperature above that of the surrounding countryside.

urbanization. The process by which large centers of population increase in size as a proportion of the total national population. In developing countries, urbanization has occurred rapidly because of migration of persons living in rural or small communities to larger urban centers of population.

urban sprawl. The patterns under which urban areas within cities expand outside of city limits and encroach upon neighboring land in response to population increases. See *conurbation.*

Uruguay Round, The. The last of eight rounds of talks held under the auspices of GATT that focused on the removal of tariffs, the development of more liberalized trade rules, and the formation of the World Trade Organization. The round began in 1986 at Punta del Este in Uruguay and was concluded in 1994 with the Marrakesh Agreement. See *Marrakesh Agreement, The.*

US Bureau of Land Management (BLM). A part of the US Department of the Interior that manages over 100 million hectares of federal land, primarily in the western states.

US Bureau of Reclamation. A US federal agency charged with water management in the western states. It also provides hydroelectric power and is a major supplier of water.

US–Canada Air Quality Agreement. A 1991 agreement that committed the two countries to reductions in sulfur dioxide and nitrogen oxide emissions, and provided a forum for addressing transboundary air issues.

user cost. The future cost, in terms of increased extraction costs in the future, and lost market opportunities from extracting an extra unit of a resource.

FURTHER READING
Randall (1987).

US National Wildlife Refuge System. A US system encompassing some 38 million hectares and 520 refuges for the protection and conservation of wildlife.

use value. From an individual's perspective, the willingness to pay for the *in situ* use of a natural resource or an environmental good.

usufruct rights. Property rights to use land or natural resources that are contingent on its use. Usufruct rights and tenure often exist with community rights where defined tracts of land may be reserved for the use of particular families or clans, but the land remains in the control of the community. Thus, upon cessation of the use of the land by a family or clan, the land may be assigned to other members of the community.

utilitarianism. A term that operationalizes the concept of the nineteenth-century economist and philosopher, Jeremy Bentham, who argued that decisions should be made to maximize the greatest good for the greatest number. A social welfare function which maximizes aggregate utility over all individuals is called a utilitarian social welfare function. See *social welfare function*.

utility. The well-being, or satisfaction, associated with an activity. Utility theory forms the basis of many economic models of consumer behavior.

utility function. A relationship where utility is a function of the quantities consumed of goods and services, and other factors.

V

value added.   The difference between the value of a good after a production process, less the value of all the inputs in producing the good.

value added tax (VAT).   A tax on the value added from one stage of the production and distribution process to another that is imposed as a percentage of the price of the good or service. Manufacturers can claim a credit on all items they purchased equal to the VAT they paid, but must pay the VAT on the goods they sell.

variable.   Anything that can vary. In mathematical problems, a variable is usually denoted by a letter or symbol.

variable costs.   Costs that vary with the level of output.

variable factor.   A factor of production that can be varied in the short-run, such as the amount of electricity.

variable rate pricing.   Pricing where the price paid per unit depends on the units used.

variance.   A measure of dispersion of a variable, defined as the average of the squared deviations from the arithmetic mean.

variance–covariance matrix.   A symmetric matrix where the elements along the main diagonal (a_{ij} where $i=j$) are the variances of the variables, and the covariances of variables i and j are given by the element a_{ij} , where i does not equal j.

vector.   1. A matrix consisting of one row or column. 2. Animals, such as insects, that transmit disease from one host to another.

vehicle bias.   The potential bias that may arise from the hypothetical method of payment in surveys designed to elicit the value that individuals may have for aspects of the environment. For example, a survey that elicits values by asking respondents if they would accept higher property taxes may yield different estimates of value than a survey that elicits values by proposing the use of higher user fees.

veld.   Open country with sparse shrubs or trees, found in South Africa.

Venturi effect. The increased velocity of a fluid or gas as it passes through a restriction in its flow. In some cities, the construction of a large buildings can create a Venturi effect with very high winds at particular places, such as street corners.

veratoxin. A toxic substance produced by some types of Escherichia coli, including the deadly Escherichia coli O157:H7. See *Escherichia coli O157:H7*.

vertical equity. The notion that people should be treated differently if they face different circumstances. The principle is applied in most countries that have a progressive income tax rate where individuals at higher incomes pay proportionately more in income tax than people at low incomes. See *horizontal equity*.

vertical integration. The merger of firms at different stages of production.

vertisol. A black soil rich in montmorillonite and that is prone to shrinking and swelling, and thus cracking.

vicarious consumption. The enjoyment of a good or service through indirect means (e.g., by viewing a film or reading literature), or through consumption of the good by others.

Vickrey auction. Auction named after the Nobel Laureate in economics, William Vickrey, that is intended to elicit truthful bids about the value of the asset or item being sold.

Vienna Convention. The document agreed to by 20 countries in 1985 through which the parties committed themselves to address what, at the time, was perceived to be a potential problem of stratospheric ozone depletion due to the emissions of chlorofluorocarbons (CFCs).

virus. A micro-organism that consists of genetic material covered in protein that reproduces inside the cells which it infects. Viruses are responsible for many diseases, including AIDS.

virtual population analysis (VPA). A method for calculating the size of a population that is frequently applied in fisheries. It uses estimates of current harvests and natural mortality to reconstruct past levels of the resource.

FURTHER READING
Hilborn and Walters (1992).

visceral characteristics. Non-scientific characteristics of animal species that influence the perceived need to preserve or protect a species.

FURTHER READING
Metrick and Weitzman (1996).

viscosity. A measure of the ability of liquids to flow, commonly measured at the speed at which a liquid flows through a tube.

volatile organic compounds (VOCs). Reactive organic compounds that can be released from the vaporization and combustion of hydrocarbons. VOCs can act as a catalyst in the production of ground-level or tropospheric ozone, which can be harmful to plants and to animals. See *ozone*.

voluntary agreements. Agreements reached between two parties based on mutually agreed-to responsibilites and obligations. Voluntary agreements are achieved without any legal sanctions or enforcement in the case of non-compliance.

voluntary risk. Risks individuals undertake on their own volition, such as crossing the road. See *involuntary risk*.

von Bertalanffy equations. Equations that relate the length (or weight) of an individual in a population to a theoretical maximum length (or weight) defined as L_∞, current age defined by t, its age at zero length (or weight) defined by t_0, and a growth parameter defined by K. Thus, the equation for growth in terms of length at time t is
$$L_t = L_\infty \left[1 - e^{-K(t-t0)} \right].$$
FURTHER READING
King (1995).

von Stackelberg duopoly. A duopoly where there exists a market leader that decides upon its production or price and, in turn, the other firm or follower bases its quantity or price decision upon this.

vulnerability. The extent to which an environment or species is threatened or endangered by changes.

vulnerability premium. The financial risk associated with a reliance on limited sources of supply of resources, and the possibility of disruption of those supplies.

W

wadi. A valley that remains dry except after heavy rainfall, at which time it will temporarily support vegetation.

wait-and-see scenario. An approach to environmental problems where, until firm evidence of a problem exists, no action is taken. The approach contrasts to the widely accepted precautionary principle. See *precautionary principle.*

Walkerton. A town in Ontario, Canada, the drinking water system of which was contaminated by *Escherichia coli* O157:H7 in May 2000. The contamination came from cow manure near a well that supplied water to the community and resulted in the death of seven individuals as well as making more than 2,300 persons seriously ill. See *Escherichia coli* O157:H7.

Walras' law. A fundamental law of economics, named after Léon Walras, that is used in general equilibrium analysis in economics. The law states that if there are n outputs (x_i) with n prices (p_i) in an economy produced by m factors (v_j) with m factor prices (w_j) there exist only $2n + 2m - 1$ independent equations necessary to find the equilibrium. In other words, the fact that the sum of $p_i x_i$ must equal the sum of $w_j v_j$ can be used to solve for one of the unknown outputs.

FURTHER READING
Takayama (1985).

warm glow. The notion that site specific valuations of a public resource or environmental good may actually be a surrogate for values given to environmental quality in general. Such values may be a potential source of bias in contingent valuation studies.

waste-end taxes. Taxes paid on the waste discharged into the environment created by the production process. Waste end taxes may be distinguished by the volume or quality of waste discharged.

water (H_2O). One of the most abundant molecules on earth that consists of one oxygen and two hydrogen atoms. Most of the earth's water is found in the oceans (97 percent). The remaining water is so-called fresh water. Over three quarters of the planet's fresh water is found in snow and ice, almost a quarter is groundwater and less than 1 percent of the total fresh water is surface water found in lakes and rivers.

water budget.   A measure of the amount of water found within an ecosystem, which takes into account how much water is coming in, how much is stored, and how much is exiting the system.

water cycle.   See *hydrological cycle.*

water pathogens.   Potentially harmful organisms, usually microbes, that exist naturally or are introduced into bathing water (e.g., ocean, seas, lakes, etc.) and drinking water.

water rights market.   A market where water rights, or the rights to access and use specified volumes of water for designated periods of time, are traded.

watershed.   An area that is the drainage basin of a defined stream or river.

watershed management.   The management of the flows, discharges and withdraws into water sources in a defined watershed so as to achieve environmental and socioeconomic objectives.

water table.   The water surface of groundwater.

Watt (W).   Unit of power that equals one joule per second.

wattle jungle.   In Swaziland, an overgrown and abandoned, but previously managed plantation of *Acacia mearnsii.*

weak sustainability.   The notion that the environment or natural capital is, to some extent, a substitute for human-made capital. See *strong sustainability.*

FURTHER READING
Neumayer (1999).

wealth.   A measure of the current value of the assets of an individual, community or society. Wealth may include tangibles, such as household goods and bank balances, as well the natural capital stock and the human capital stock.

weather.   At a point in time and place, the state of the atmosphere as measured by its temperature, wind speed, precipitation, humidity, and other factors.

weighted average.   An average value determined by weights. For observations x_1, x_2, ... x_n and weights w_1, w_2, ... w_n, it is calculated as $(w_1 x_1 + w_2 x_2 + ... + w_n x_n)/(w_1 + w_2 + ... + w_n)$.

welfare aggregation.   The addition of a measure of the economic well-being of individuals to construct a measure of the overall well-being of society.

welfare change.   A measure of the change in the well-being of an individual as a result of a change in some state of the world.   See *compensating surplus variation, equivalent variation,* and *consumer surplus*.

welfare economics.   The field of economics that investigates the interaction between the economic well-being of individuals and economic efficiency.

Wellman–Lord process.   A highly effective method for removing sulfur dioxide from the emissions of coal-burning generators.

wet biomass.   The weight of living organisms at an instant in time within a given area, including any water found within the organism.

wet deposition.   The transmission of air pollutants from the atmosphere to the earth's surface via precipitation.

wetland pollution filter.   The use of wetlands, either natural or artificial, to convert effluent and liquid waste into clean water.

wetlands.   Areas of land characterized by the presence of either standing or flowing water for at least a portion of the year, including marshes, swamps, bogs, shallow tidal areas, and ground that is waterlogged. Wetlands are important habitats for fauna, and staging areas for migrating birds.

white noise series.   A series of independent random variables that are independently and identically distributed with zero mean.

wilderness. The 1964 United States Wilderness Act defines wilderness as "an area where the earth and its community of life are untrammelled by man, where man himself is the visitor that does not remain". More generally, wilderness is used to describe natural habitats that are largely free from human activity or development.

wilderness tourism. A subset of "ecotourism" in which destinations are primarily in remote, wilderness areas that are largely free from any development.

wildlife corridor. Linear areas used by wildlife to move between different habitats (e.g., river valleys used by elk to reach their summer range in the mountains and then return to their winter range in the foothills).

wildlife refuges. Locations where wildlife is protected for conservation purposes. The USA currently has 36 million hectares designated as part of its national wildlife refuge system.

willingness to accept. The compensation required to return an individual to his or her original state of economic well-being following some change (possibly hypothetical) in the world. In many cases, a willingness to accept measure is considered to be inferior to the willingness to pay measure of welfare change.

willingness to pay. The willingness of an individual to pay (i.e., give up part of one's budget) in order to secure a good or service. Willingness to pay is often used to refer to the amount that a consumer would pay for a hypothetical good, service, or change in some state of the world, particularly as described in a contingent valuation survey.

wind chill. The cooling of a body via wind which draws away latent heat. The wind-chill index measures how much body heat is lost per square meter of skin per hour and provides an equivalent (and colder) temperature that would be comparable if there were no wind.

windrow. The accumulation of slash or other vegetative debris on a forest site that has been cut for harvest. The windrow usually is intended to promote regeneration at that site.

windthrow. See *blowdown*.

wise use movement.   A movement of various individuals, groups and communities that supports the development of natural resources. One of the goals of the movement is to ensure the primacy of private-property rights in the regulation of the environment, and to oppose large-scale government intervention.

withdrawal permits.   Permits that allow the holder to use a given volume of water for a specified period of time.

withdrawal rate.   Rate at which water is removed from an aquifer.

withdrawal right.   A property right that enables the holder to extract or harvest a resource, such as a hunting permit.

withholding period.   The period of time a food product must be withheld from sale or consumption after a treatment or exposure to a toxic substance. Many fruits have a withholding period after they have been sprayed with pesticides.

woodlot.   The forested area of privately-owned land used for harvesting trees.

World Bank.   See *International Bank for Reconstruction and Development.*

World Health Organization (WHO).   A United Nations agency formed in 1948 and mandated with improving the health of the world's people.

World Trade Organization (WTO).   The successor to the GATT that was established in 1995, is mandated to expand multilateral trade. WTO has a dispute resolution mechanism to help prevent restrictive trade practices.

FURTHER READING
International Institute for Sustainable Development (2000).

X

X-efficiency. Term coined by H. Leibenstein to describe the effectiveness of managers at minimizing the costs of production for a given level of output.

xerophyte. Plant adapted to dry or desert conditions, such as a cactus.

Y

yard.   Unit of length equal to 0.91 meters.

year-class.   A method of defining cohorts within a population based upon the year that they were born or were spawned.

year-class strength.   Term that refers to the size of a cohort in a population relative to older and younger cohorts. See *cohort*.

yellowcake.   Name given to a mixture of uranium that is used as a fuel in nuclear reactors.

yellow fever.   A life-threatening disease caused by a virus transmitted by mosquitoes.

younger dryas.   A period of sudden cold temperatures in the northern hemisphere that began about 10,000 to 11,000 years ago and which occurred after the end of the last ice age. A hypothesis is that the sudden cooling may have arisen from the rapid melting of Greenland glaciers which, in turn, created a cold mass of water that diverted the Gulf Stream from the North Atlantic.

Z

zero discharge.   No measurable emissions from a defined source.

zero emissions.   See *zero discharge.*

zero population growth.   An often-stated goal of some environmentalists that the human population of the world be stabilized at its current level.

zero-revenue auction.   An auction designed to allocate assets or property rights among firms or users where the revenue collected from the auction is subsequently returned to owners of the newly acquired assets.

zero sum game.   A mathematical construction of a set of choices where the outcomes are symmetrically opposed such that the total benefits received by one party are equivalent to the losses borne by the other. See *non-zero sum game.*

zero wealth effects.   An assumption made under the Coase Theorem that, in the absence of property rights, creating and giving a property right to any one party does not affect the wealth of any of the parties, and thus does not affect the ability of the parties to reach the efficient outcome. See *Coase Theorem.*

zoning.   Collective term for regulations that govern the permitted land use in defined areas. Zoning can be used to conserve green spaces within urban environments and to prevent industries from setting up their operations in residential neighborhoods.

zoogeography.   The study of the spatial distribution of animals.

zoology.   The study of animal life.

zooplankton.   Plankton consisting of small creatures that include the juvenile stages of many larger aquatic creatures. See *plankton* and *phytoplankton.*

References

Australian Bureau of Agricultural and Resource Economics (1997), *The Economic Impact of International Climate Change Policy*, Canberra: Commonwealth of Australia.

Abrams, P.A. (1995), 'Monotonic or Unimodal Diversity–Productivity Gradients: What Does Competition Theory Predict?', *Ecology*, October, **76(7)**, 2019-2027.

Ahrens, C.D. (2000), *Meteorology Today: An Introduction to Weather, Climate, and the Environment*, Pacific Grove, California: Brooks/Cole.

Akerlof, G. (1970), 'The Market for "Lemons": Quality, Uncertainty and the Market Mechanisms', *Quarterly Journal of Economics*, **84**, 488-500.

Allaby, M. (1996), *Basics of Environmental Science*, London: Routledge.

Anderson, M.S. (1994), *Governance by Green Taxes: Making Pollution Prevention Pay*, Manchester: Manchester University Press.

Anderson, S.H. (1999), *Managing Our Wildlife Resources*, Upper Saddle River, New Jersey: Prentice Hall.

Arnold, J.E.M. and J.G Campbell (1986), 'Collective Management of Hill Forests in Nepal: the Community Forest Development Project', in *Proceedings on the Conference on Common Property Management*, National Academy Press, 425-454.

Arrow, K. and A. Fisher (1974), 'Environmental Preservation, Uncertainty, and Irreversibility', *Quarterly Journal of Economics*, **88(1)**, 312-319.

Barbier, E.B., (ed.) (1993), *Economics and Ecology*, London: Chapman and Hall.

Barro, R.J. and X. Sala-i-Martin (1995), *Economic Growth*, New York: McGraw-Hill Inc.

Barry, R.G. and R.J. Chorley (1992), *Atmosphere, Weather and Climate*, 6th Edn, London: Routledge.

Baumol, W.J. and W.E. Oates (1988), *The Theory of Environmental Policy*, 2nd Edn, Cambridge: Cambridge University Press.

Bierman, H.S. and L. Fernandez (1993), *Game Theory with Economic Applications*, Reading, MA: Addison-Wesley Publishing.

Binmore, K. (1992), *Fun and Games: A Text on Game Theory*, Lexington, MA: D.C. Heath and Company.

Bloom, A.L. (2001), *Geomorphology: A Systematic Analysis of Late Cenozoic Landforms*, Upper Saddle River, New Jersey: Prentice Hall.

Boserup, E. (1965), *The Conditions of Agricultural Growth*, London: Allen Unwin.

Brennan, A. (1992), 'Moral Pluralism and the Environment', *Environmental Values*, 1, 15-33.

Bromley, D.W. (1991), *Environment and Economy: Property Rights and Public Policy*, Oxford: Basil Blackwell.

Brown, G. and R. Mendelsohn (1984), 'The Hedonic Travel Cost Method', *Review of Economics and Statistics*, 66, 427-433.

Brown, J.H., D.W. Mehlman and G.C. Stevens (1995), 'Spatial Variation in Abundance', *Ecology*, October, 76(7), 2028-2043.

Brubaker, S. (ed.) (1984), *Rethinking the Federal Lands*, Washington, DC: Resources for the Future, Inc.

Bustamante, R.H., G.M. Branch and S. Eekhout (1995), 'Maintenance of an Exceptional Intertidal Grazer BioMass in South Africa', *Ecology*, October, 76(7), 2314-2329.

Callan, S.J. and J.M. Thomas (2000), *Environmental Economics and Management: Theory, Policy, and Applications*, Orlando, Florida: The Dryden Press.

Campbell, N.A., Reece, J.B. and L.G. Mitchell (1999), *Biology*, 5th Edn., Menlo Park: Addison-Wesley Longman.

Carson, R. (1962), *Silent Spring*, New York; Houghton Mifflin.

Carson, R.T., N.E. Flores, K.M. Martin and J.L. Wright (1996), 'Contingent Valuation and Revealed Preference', *Land Economics*, February, **72(1)**, 80-99.

Case, K.E. & R.C. Fair (1999), *Principles of Economics*, Upper Saddle River, New Jersey: Prentice Hall.

Chiang, A.C. (1992), *Elements of Dynamic Optimization*, New York: McGraw-Hill.

Chow, G. (1960), 'Tests of Equality between Sets of Coefficients in Two Linear Regressions', *Econometrica*, July, **28(3)**, 591-605.

Ciriacy-Wantrup, S.V. (1952), *Resource Conservation: Economics and Policy*, Berkeley: University of California Press.

Ciriacy-Wantrup, S.V. and R.C. Bishop (1975), 'Common Property as a Concept in Natural Resources Policy', *Natural Resources Journa,,* **15**, 713-727.

Clark, C. (1990), *Mathematical Bioeconomics*, 2nd Edn, New York: John Wiley & Sons.

Coase, R.H. (1960), 'The Problem of Social Cost,' *Journal of Law and Economics*, **3**, 1-44.

Cohen, J. and I. Stewart (1994), *The Collapse of Chaos*, New York: Penguin Books

Colborn, T., D. Dumanoski and J.P. Myers (1997), *Our Stolen Future*, New York: Plume/Penguin.

Conrad, J. (1995), 'Bioeconomic Models of the Fishery', in *The Handbook of Environmental Economics*, Cambridge, MA: Basil Blackwell.

Cornes, R. and T. Sandler (1996), *The Theory of Externalitie,, Public Goods and Club Goods*, 2nd Edn, Cambridge University Press.

Costanza, R., O. Segura and J. Martinez-Alier (1996), *Getting Down to Earth*, Washington, DC: Island Press.

Costanza, R., F. Sklar and M. White (1990), 'Modeling Coastal Landscape Dynamics', *BioScience*, **40**, 91-97.

Cropper, M. and Sussman F.G. (1990), 'Valuing Future Risks To Life', *Journal of Environmental Economics and Management*, **20(2)**, 160-174.

Daly, H.E. (1973), 'The Steady-State Economy: Toward a Political Economy of Biophysical Equilibrium and Moral Growth', in H.E. Daly (ed.), *Toward a Steady-State Economy*, San Francisco: W.H. Freeman and Company.

Daly, H.E. (1996), 'Consumption: Value Added, Physical Transformation, and Welfare', in R. Costanza, O. Segura and J. Martinez-Alier (eds)., *Getting Down to Earth*, Washington, DC: Island Press, 49-59.

Darwin, C. (1859), *On the Origin of Species by means of Natural Selection or Preservation of Favoured Races in the Struggle for Life*, 1st Edn, London: Murray

Davis. G.A. (1996), 'Option Premiums in Mineral Asset Pricing', *Land Economics*, May, **72(2)**, 167-186.

Davis, P. (1999), *The Fifth Miracle: The Search for the Origin and Meaning of Life*, New York: Simon Schuster.

Dawkins, R. (1976), *The Selfish Gene*, Oxford: Oxford University Press.

Deaton, M.L. and J.J. Winebrake (2000), *Dynamic Modeling of Environmental Systems*, New York: Springer-Verlag.

de Bruyn, S.M. (2000), *Economic Growth and the Environment: An Empirical Analysis*, Dordrecht: Kluwer Academic Publishers.

Devall, B. and G. Sessions (1985), *Deep Ecology*, Salt Lake City, Utah: Peregrine Smith.

Devlin, R.A. and R.Q. Grafton (1998), *Economic Rights and Environmental Wrongs: Property Rights for the Common Good*, Cheltenham: Edward Elgar.

Dixit, A. and V. Norman (1980), *International Trade: Theory of International Trade: a Dual, General Equilibrium Approach*, Welwyn, Herts: J. Nisbet.

Dotto, L. (1986), *Planet Earth in Jeopardy: Environmental Consequences of Nuclear War*, Chichester: John Wiley and Sons.

Duraiappah, A.K. (1993), *Global Warming and Economic Development*, Boston: Kluwer Academic Publishers.

Eggertsson, T. (1990), *Economic Behavior and* Institution,s Cambridge, UK: Cambridge University Press.

Ekins, P. (2000), *Economic Growth and Environmental Sustainability: The Prospects for Green Growt,,* London: Routledge.

Fastie, C.L. (1995), 'Causes and Ecosystem Consequences of Multiple Pathways of Primary Succession at Glacier Bay, Alaska', *Ecology*, **76(6)**, 1899-1916.

Feeny, D., S. Hanna and A.F. McEvoy (1996), 'Questioning the "Tragedy of the Commons" Model of Fisheries', *Land Economics*, May, **72(2)**, 187-205.

Fetter, C.W. (2001), *Applied Hydrogeology*, Upper Saddle River, New Jersey: Prentice Hall.

Field, B.C. (2000), *Natural Resource Economics: An Introduction*, Boston, MA: McGraw-Hill.

Fisher, A. and R. Raucher (1984), 'Instrinsic Benefits of Improved Water Quality: Conceptual and Empirical Perspectives', in V. Kerry Smith and Ann Dryden Witte (eds)., *Advances in Applied Microeconomics*, Greenwich, CT: JAI Press.

Ford, A. (1999), *Modeling the Environment: An Introduction to System Dynamics Modeling of Environmental Systems*, Washington DC: Island Press.

Forest, B.C. (2001) www.for.gov.bc.ca/pab/publctns/glossary/glossary.htm

Fox, N. (1998), *Spoiled: Why Our Food is making Us Sick and What We Can Do about It*, New York: Penguin Books.

Fox, W. (1990), *Toward a Transpersonal Ecology*, Boston: Shambhala.

Frank-Kamenetskii, M.D. (1997), *Unraveling DNA: The Most Important Molecule of Life*, Reading, MA: Addison-Wesley.

Freeman, S. and J.C. Herron (2000), *Evolutionary Analysis*, Upper Saddle River, New Jersey: Prentice Hall.

Gale, R.P. and S.M. Cordray (1994), 'Making Sense of Sustainability: nine answers to "What Should Be Sustained"', *Rural Sociology*, **59(3)**, 311-332.

Ganzhorn, J.G. (1995), 'Low Level Forest Disturbance Effects on Primary Production, Leaf Chemistry and Lemur Populations', *Ecology*, October, **76(7)**, 2084-2096.

Garnaut, R. and A.C. Ross (1975), 'The Neutrality of the Resource Rent Tax', *The Economic Record*, **55**, 193-201.

Garrod, G. and K.G. Willis (1999), *Economic Valuation of the Environment: Methods and Case Studies*, Northampton, MA: Edward Elgar.

Georgescu-Roegen, N. (1973), 'The Entropy Law and the Economic Problem', in H.E. Daly (ed.), *Toward a Steady-State Economy*, San Francisco: W.H. Freeman and Company.

Gillis, M., D. Perkins, M. Roemer and D. Snodgrass (1997), *Economics of Development*, 5th Edn, New York: W.W. Norton.

Gilroy, J.M. (ed.) (1993), *Environmental Risk, Environmental Values, and Political Choices*, Boulder: Westview Press.

Gleick, J. (1987), *Chaos: making a new science*, Fairfield, PA: Arcata Graphics.

Goldberger, A.S. (1964), *Econometric Theory*, New York: Wiley.

Gould, S.J. (1989), *Wonderful Life: The Burgess Shale and the Nature of History*, New York: W.W. Norton.

Gould, S.J. (1993), 'Reconstructing (and Deconstructing) the Past', in S.J. Gould, (ed.), *The Book of Life*, Toronto, Canada: Viking.

Grafton, R.Q. (1996), 'Individual Transferable Quotas: Theory and Practice', *Reviews in Fish Biology and Fisheries*, **6**, 5-20.

Grafton, R.Q. (2000), 'Governance of the Commons: A Role for the State?', *Land Economics*, November, **76(4)**, 504-517.

Grafton, R.Q. and R.A. Devlin (1996), 'Paying for Pollution: Permits and Charges', *Scandinavian Journal of Economics*, **98**, 275-288.

Grafton, R.Q. and H.W. Nelson (1997), 'Fishers' Individual Salmon Harvesting Rights: An Option for Canada's Pacific Fisheries', *Canadian Journal of Fisheries and Aquatic Sciences* **54**, 474-482.

Grafton, R.Q., L.K. Sandal and S.I. Steinshamn (2000), 'How to Improve the Management of Renewable Resources: The Case of Canada's Northern Cod Fishery', *American Journal of Agricultural Economics*, **82**, 570-580.

Grafton, R.Q. and T.C. Sargent (1997), *A Workbook in Mathematical Economics for Economists*, New York: McGraw-Hill.

Grafton, R.Q. and J. Silva-Echenique (1997), 'How to Manage Nature? Strategies, Predator–Prey Models, and Chaos', *Marine Resource Economics*, **12**, 127-143.

Grafton, R.Q., D. Squires and K.J. Fox (2000), 'Private Property and Economic Efficiency: A Study of a Common-Pool Resource', *The Journal of Law and Economics*, **43(2)**, 679-713.

Granger, C.W.J. (1969), 'Investigating Causal Relations by Econometric Models and Cross-Spectral Models', *Econometrica*, **37**, 424-438.

Greene, W.H. (1997), *Econometric Analysis,* Upper Saddle River, New Jersey: Prentice-Hall.

Griffiths, A.J.F. (1999), *Modern Genetic Analysis*, New York: W.H. Freeman.

Grifo, F.T. (1999), 'Infectious Diseases and the Loss of Biodiversity through Deforestation', in R. DeSalle (ed.), *Epidemic! The World of Infectious Diseases*, New York: The New Press in conjunction with The American Museum of Natural History.

Griliches, Z. (1971), *Price Indexes and Quality Change,* Cambridge, MA: Harvard University Press.

Grossman, M. (1972), 'On the Concept of Health Capital and the Demand for Health', *Journal of Political Economy*, **80(2)**, 223-255.

Guttorp, P. (1995), *Stochastic Modelling of Scientific Data*, London: Chapman and Hall.

Hanley, N., J.F. Shogren, and B. White (1997), *Environmental Economics in Theory and Practice,* New York: Oxford University Press.

Hardie, I.W. and P.J. Parks (1996), 'Reforestation Cost-Sharing Programs', *Land Economics*, May, **72(2)**, 248-260.

Hardin, G. (1968), 'The Tragedy of the Commons', *Scienc*, **162**, 1143-1248.

Harris, D.P. and B.J. Skinner (1982), 'The Assessment of Long-term Supplies of Minerals', in V.K. Smith and J.V. Krutilla (eds.), *Explorations in Natural Resource Economics*, The Johns Hopkins University Press.

Hartman, R. (1976), 'The Harvesting Decision When a Standing Forest Has Value', *Economic Inquiry*, **14**, 52-58.

Hartwick, J.M. (1977), 'Intergenerational Equity and the Investing of Rents from Exhaustible Resources', *American Economic Review*, **67**, 972-974.

Hartwick, J.M. and N. Olewiler (1998), *The Economics of Natural Resource Use*, Reading, MA: Addison-Wesley.

Hazilla, M. and R.J. Kopp (1990), 'The Social Cost of Environmental Quality Regulations: A General Equilibrium Analysis', *Journal of Political Economy*, **98(4)**, 853-873.

Hilary, S. (1996), 'Cross-Media Pollution', *Land Economics*, August, **72(3)**, 298-312.

Hill, C., W. Griffiths and G. Judge (1998), *Undergraduate Econometrics*, New York: John Wiley & Sons.

Hilborn, R. and C.J. Walters (1992), *Quantitative Fisheries Stock Assessment: Choice, Dynamics and Uncertainty,* New York: Chapman and Hall.

Hobson, A. (1999), *Physics Concepts and Applications*, Upper Saddle River, New Jersey: Prentice Hall.

Holling, C.S. (1973), 'Resilience and Stability of Ecosystems', *Annual Review of Ecology and Systematics*, **4**, 1-23.

Holmberg, J., K. Robert and K. Erikson (1996), 'Socio-Ecological Principles for a Sustainable Society', in R. Costanza, O. Segura and J. Martinez-Alier (eds.), *Getting Down to Earth*, Washington, DC: Island Press, 17-48.

Houghton, J. (1997), *Global Warming: The Complete Briefin,,* 2nd Edn, Cambridge, UK: Cambridge University Press.

Howe, H.F. (1995), 'Succession and Fire Season in Experimental Prairie Plantings', *Ecology*, **76(6)**, 1917-1925.

Hufbauer, G.C. and J.J. Schott (1998), 'North American Economic Integration: 25 Years Backward and Forward', Industry Canada Research Publication Program.

Intergovernmental Panel on Climate Change (1997), 'An Introduction to Simple Climate Models Used in the IPCC', *Second Assessment Report*, UN Environmental Program.

Intergovernmental Panel on Climate Change (1997), *Stabilization of Atmospheric Greenhouse Gases: Physical, Biological and Socio-Economic Implications*, Geneva: United Nations Environmental Program.

International Institute for Sustainable Development (2000), *Environment and Trade: A Handbook*, Winnipeg, Manitoba: International Institute for Sustainable Development.

Johnson, S.L. and D.M. Pekelney (1996), 'Regional Clean Air Incentives Market', *Land Economics*, August, **72(3)**, 277-297.

Katz, M.L. and H.S. Rosen (1998), *Microeconomics*, 3rd Edn, Boston, MA: Irwin McGraw-Hill.

Kaufman, L. and K. Mallory (1993), *The Last Extinction*, Cambridge, MA: MIT Press.

Kennedy, P. (1998), *A Guide to Econometrics*, 4th Edn, Oxford: Basil Blackwell.

Keynes, J.M. (1936), *The General Theory of Employment, Interest, and Money*, London: Macmillan.

Kimmins, J.P. (1997), *Forest Ecology: A Foundation for Sustainable Management*, 2nd Edn, Upper Saddle River, New Jersey: Prentice Hall.

King, M. (1995), *Fisheries Biology, Assessment and Management*, Osney Mead, England: Fishing News Books.

Krebs, C.M. (1994), *Ecology*, New York: HarperCollins.

Krugman, P. and M. Obstfeld (1997), *International Economics: Theory and Policy*, Reading, MA: Addison-Wesley Longman.

Krutilla, J.V. and A.C. Fisher (1975), *The Economics of Natural Environments: Studies in the Valuation of Commodity and Amenity Resources*, Baltimore, MD: The Johns Hopkins University Press.

Laughland, A.S., W.N. Musser, J.S. Shortle and L.M. Musser (1996), 'Averting Cost Measures of Environmental Benefits', *Land Economics*, February, **(72)1**, 100-112.

Libecap, G.D. (1989), *Contracting for Property Rights*, New York: Cambridge University Press.

Lipsey, R.G. and K. Lancaster (1956), *The General Theory of the Second Best*, Cheltenham, UK: Edward Elgar.

Louviere, J. (1996), 'Relating Stated Preference Measures and Models to Choices in Real Markets: Calibration of CV Responses', in D.J. Bjornstad and J.R. Kahn (eds.), *The Contingent Valuation of Environmental Resources: Methodological Issues and Research Needs*, Cheltenham, UK: Edward Elgar.

Lovelock, J.E. (1990), 'Hands Up for the Gaia Hypothesis', *Nature*, March, **344**, 100-102.

Lovelock, J.E. (2000), *Gaia: A New Look at Life on Earth*, 2nd Edn, Oxford: Oxford University Press.

Marcouiller, D.W., D.K. Lewis and D.F. Schreiner (1996), 'Timber Production Factor Shares', *Land Economics*, August, **72(3)**, 358-369.

McConnell, K.E. (1983), 'Existence and Bequest Value', in R.D. Rowe and L.G. Chestnut, (eds.), *Managing Air Quality and Scenic Resources at National Parks and Wilderness Areas*, Boulder: Westview Press.

McConnell, K.E. (1990), 'Models for Referendum Data: the Structure of Discrete Choice Models for Contingent Valuation', *Journal of Environmental Economics and Management*, **18(1)**, 19-34.

Meadows, D.H., D.L. Meadows, J. Randers and W.W. Behrens III (1974), *The Limits to Growth*, 2nd Edn, New York: Signet.

Metrick, A. and M.L. Weitzman (1996), 'Endangered Species Preservation', *Land Economics*, February, **72(1)**, 1-16.

Millington, A.C., T.D. Douglas and P. Ryan (1994), 'Glossary' in *Estimating Woody Biomass in Sub-Saharan Africa*, Washington, DC: World Bank, 175-178.

Morris, P. and R. Therivel (1995), *Methods of Environmental Impact Assessment*, Vancouver: UBC Press.

Mukherjee, C., H. White and M. Wuyts (1998), *Econometrics and Data Analysis for Developing Countries*, New York: Routledge.

National Geographic (March 1999), 'Ten Years After *Exxon Valdez*', **195(3)**, 98-117.

National Research Council (1999), *Sharing the Fish: Toward a National Policy on Individual Fishing Quotas*, Washington, DC: National Academy Press.

Nelson (1995), *Public Lands and Private Rights*, Lanham, MD: Rowman and Littlefield.

Neumayer, E. (1999), *Weak Versus Strong Sustainability: Exploring the Limits of Two Opposing Paradigms*, Cheltenham, UK: Edward Elgar.

Nordhaus, W.D. and E.C. Kokkelenberg (1999), *Nature's Numbers: Expanding the National Economic Accounts to Include the Environment*, Washington, DC: National Academy Press.

O'Brien, S.T., S.P. Hubbell, P. Spiro, R. Condit and R.B. Foster (1995), 'Diameter, Height, Crown and Age Relationships in Eight Neotropical Tree Species', *Ecology*, **76(6)**, 1926-1939.

Organization for Economic Co-operation and Development (1999), *Trade Measures in Multilateral Environmental Agreements*, Paris, France: OECD Publications.

Ostrom, E. (1990), *Governing the Commons: The Evolution of Institutions for Collective Action*, Cambridge, MA: Cambridge University Press.

Ostrom, E., R. Gardner and J. Walker (1994), *Rules, Games and Common-Pool Resources*, Ann Arbor: The University of Michigan Press.

Park, C.C. (1992), *Tropical Rainforests*, London: Routledge.

Perman, R., Y. Ma and J. McGilvray (1996), *Natural Resource and Environmental Economics*, London: Longman.

Perrings, C. (1991), 'Reserved Rationality and the Precautionary Principle: Technological Change, Time and Uncertainty in Environmental Decision making', in R. Costanza (ed.), *Ecological Economics: the Science and Management of Sustainability*, New York: Columbia University Press, 153-66.

Phillips, A.W. (1958), 'The Relationship between Unemployment and the Rate of Change of Money Wages in the United Kingdom, 1861–1957', *Economica*, **25**, 283-299.

Pielou, E.C. (1998), *Fresh Water*, Chicago: University of Chicago Press.

Postgate, J. (1992), *Microbes and Man*, Cambridge, MA: Cambridge University Press.

Ramanathan, R. (1992), *Introductory Econometrics*, 2nd Edn, New York: Dryden Press.

Randall, A. (1987), *Resource Economics: An Economic Approach to Natural Resource and Environmental Policy*, New York: John Wiley & Sons.

Rao, P.K. (2000), *Sustainable Development–Economics and Policy*, Malden, MA: Basil Blackwell.

Rawls, J. (1971), *A Theory of Social Justice*, Cambridge, MA: Harvard University Press.

Ray, D. (1998), *Development Economics*, Princeton, New Jersey: Princeton University Press.

Reuben, K.U. (1982), *The Human Geography of Tropical Africa*, Ibadan: Heinemann Educational Books (Nigeria) Ltd.

Romer, D. (1996), *Advanced Macroeconomics*, New York: McGraw-Hill.

Romme, W.H., M.G. Turner, L.L. Wallace and J.S. Walker (1995), 'Aspen, Elk and Fire in Northern Yellowstone Park' *Ecology*, October, **76(7)**, 2097-2106.

Rosen, S. (1974), 'Hedonic Prices and Implicit Markets: Product Differentiation in Pure Competition' *Journal of Political Economy*, **82,** 34-55.

Ross, M.R. (1997), *Fisheries Conservation and Management*, Upper Saddle River, New Jersey: Prentice Hall.

Salisbury, F.B. and C.W. Ross (1992), *Plant Physiology*, Belmont: Wadsworth.

Schumacher, E.F. (1973), *Small is Beautiful: Economics as if People Mattered*, New York: Harper and Row.

Serageldin, I. (1995), *Toward Sustainable Management of Water Resources. Directions in Development Series*, Washington, DC: World Bank.

Sheffrin, S.M. (1983), *Rational Expectations*, Cambridge, UK: Cambridge University Press.

Shone, R. (1997), *Economic Dynamics,* Cambridge, MA: Cambridge University Press.

Slade, M.E. (1996), 'Compliance Costs for Mineral Commodities', *Land Economics*, February, **72(1)**, 17-32.

Smith, D.M., B.C. Larson, M.J. Kelty, P. Mark and S. Ashton (1997), *The Practice of Silviculture: Applied Forest Ecology*, New York: Wiley.

Smith, R.L. and T.M. Smith (2000), *Elements of Ecology*, San Francisco: Addison-Wesley Longman.

Somerville, R.C.J. (1996), *The Forgiving Air: Understanding Environmental Change*, Berkeley: University of California Press.

Stevens. J.B. (1996), 'John Locke, Environmental Property, and Instream Water Rights', *Land Economics*, May, **72(2)**, 261-268.

Stiling, P.D. (1992), *Introductory Ecology*, Englewood Cliffs, New Jersey: Prentice Hall.

Stuart, A. and K. Ord (1994), *Kendall's Advanced Theory of Statistics, Volume 1*, London: Edward Arnold.

Studenmund, A.H. (2001), *Using Econometrics: Practical Guide* 4th Edn, Boston, MA: Addison-Wesley Longman.

Suberkropp, K. and E. Cauvet (1995), 'Regulation of Leaf Breakdown by Fungi in Streams: Influences of Water Chemistry', *Ecology*, July, **76(5)**, 1433-1445.

Takayama, A. (1985), *Mathematical Economics*, 2nd Edn, Cambridge, MA: Cambridge University Press.

Thornton, J. (2000), *Pandora's Box: Chlorine, Health, and a New Environmental Strategy*, Cambridge, MA: MIT Press.

Tietenberg, T. (1994), *Environmental Economics and Policy*, New York: Harper Collins.

Tietenberg, T.H. (1996), *Environmental and Natural Resource Economics*, 4th Edn, New York: HarperCollins.

Tolba, M.K. and O.A. El-Kholy (1992), *The World Environment 1972-1992*, London: Chapman and Hall.

Tomar, M. (1999), *Quality Assessment of Water and Wastewater*, New York: Lewis Publishers.

Van Kooten, G.C. (1993), *Land Resource Economics and Sustainable Development: Economic Policies and the Common Good*, Vancouver, BC: University of British Columbia Press.

Van Kooten, G.C. and E.H. Bulte (2000), *The Economics of Nature*, Oxford: Blackwell Publishers.

Vogler, J. (2000), *The Global Commons: Environmental and Technological Governance*, 2nd Ed., Chichester, UK: John Wiley & Sons Ltd.

Walters, C.J. and R. Hilborn (1976), 'Adaptive Control of Fishing Systems', *Journal of The Fisheries Research Board of Canada*, **33**, 145-159.

Wicander, R. and J.S. Monroe (1999), *Essentials of Geology*, Pacific Grove, California: Brooks/Cole.

Wieder, R.K. and J.S. Wright (1995), 'Tropical Forest Litter Dynamics and Dry Season Irrigation on Barro Colorado Island, Panama', *Ecology*, **76(6)**, 1971-1979.

Wiesbrod, B. (1964), 'Collective Consumption Services of Individual Consumption Goods', *QJE*, **77(3)**, 71-77.

Williams, S.L. (1995), 'Surfgrass (Phyllospadix Torreyi) Reproduction: Reproductive Prenology, Resource Allocation and Male Rarity', *Ecology*, **76(6)**, 1953-1970.

Wills, C. (1989), *The Wisdom of the Genes*, New York: Basic Books.

Wilson, R.C.L., S.A. Drury and J.L. Chapman (2000), *The Great Ice Age: Climate Change and Life*, London: Routledge.

Winteringham, P.W. (1992), *Energy Use and the Environment*, London: Lewis Publishers.

World Commission on Environment and Development (1987), *Our Common Future*, Oxford: Oxford University Press.

World Commission on Forests and Sustainable Development (1999), *Our Forests, Our Future*, Cambridge, MA: Cambridge University Press.

Zerbe, R.O. and D.D. Dively (1994), *Benefit–Cost Analysis in Theory and Practice*, New York: HarperCollins.

Appendix 1: Greek Alphabet

LETTER	NAME
A α	Alpha
B β	Beta
Γ γ	Gamma
Δ δ	Delta
E ε	Epsilon
Z ζ	Zeta
H η	Eta
Θ θ	Theta
I ι	Iota
K κ	Kappa
Λ λ	Lambda
M μ	Mu
N ν	Nu
Ξ ξ	Xi
O ο	Omicron
Π π	Pi
P ρ	Rho
Σ σ	Sigma
T τ	Tau
Y υ	Upsilon
Φ φ	Phi
X χ	Chi
Ψ ψ	Psi
Ω ω	Omega

Appendix 2: Roman Numerals

I	1
II	2
III	3
IV	4
V	5
VI	6
VII	7
VIII	8
IX	9
X	10
XX	20
XXX	30
XL	40
L	50
LX	60
LXX	70
LXXX	80
XC	90
C	100
CC	200
CCC	300
CD	400
D	500
DC	600
DCC	700
DCCC	800
CM	900
M	1000

Appendix 3: Système Internationale Units

SI units	Conversions

LENGTH
Meter (m)
Centimeter (cm)
Kilometer (km)
Micrometer (μm)
	1 m = 3.2808 ft 1 ft = 0.3048 m
	1 cm = 0.3937 in 1 in = 2.54 cm
	1 km = 0.6214 mile 1 mile = 1.6093 km
	1 μm = 10^{-6} m
	1 ångstrom = 10^{-10} m
	One degree of latitude = 111.1 km
	                                    = 69.1 miles
	                                    = 60 nautical miles
	One nautical mile = 1.15 statute miles
	                                    = 1.85 km

MASS Kilogram (kg) Gram (g) Metric tonne (t)	1 kg = 2.2046 lb 1 lb = 0.4536 kg 1 g = 0.0353 oz 1 oz = 28.3495 g 1 tonne =10^3 kg = 1.10 short ton = 2204.622 lbs
TEMPERATURE Kelvin (K)	1° K = 1° C = 1.8° F 1° F = 5.9° C = 5.9° K ° C + 273.15 = K ° C × (1.8) + 32 = ° F (° F - 32) × (5/9) = ° C
AREA Square meter (m²) Square centimeter (cm²) Square kilometer (km²) Hectare (ha)	1 m² = 10.764 sq ft 1 sq ft = 0.0929 m² 1 cm² = 0.155 sq in 1 sq in = 6.4516 cm² 1 km² = 0.3861 sq miles 1 sq mile = 2.5899 km² 1 ha = 10 000 m² = 2.471 acres 1 acre = 43 560 sq ft = 0.4047 ha

VOLUME	

Cubic meter (m³)
 	
Cubic centimeter (cm³)
	1 m³ = 35.3 cu ft
	1 cu ft = 0.028 m³
	1 cm³ = 0.061 cu in
	1cu in = 16.39 cm³
	1 liter (l) = 10^{-3}m³
	            = 0.264 gal (US)
	            = 0.22 gal (Imperial)
	1 barrel = 158.983 litres

WORK AND ENERGY Joule (J)	1 calorie (cal) = 4.186 J 1 ft lb force per sec = 1.36 J per sec 1 J = 0.738 ft lb force per sec 1 BTU = 1055 J
POWER Watt (W)	1 W = 1 J per sec 1 horse power (hp) = 33 000 ft lb per min = 746 W

Appendix 4 : Prefixes of the Système Internationale (SI)

Prefix	Abbreviation	Multiplication factor
exa	E	10^{18}
peta	P	10^{15}
tera	T	10^{12}
giga	G	10^{9}
mega	M	10^{6}
kilo	k	10^{3}
hecto	h	10^{2}
deca	da	10^{1}
deci	d	10^{-1}
centi	c	10^{-2}
milli	m	10^{-3}
micro	μ	10^{-6}
nano	n	10^{-9}
pico	p	10^{-12}
femto	f	10^{-15}
atto	a	10^{-18}

Appendix 5 : Common Abbreviations

A	ampere
Bq	becquerel
cm	centimeter
°C	degree Celsius
d	day
dB	decibel
g	gram
GWh	gigawatt hour
ha	hectare
h	hour
Hz	hertz
J	joule
kg	kilogram
km	Kilometer
km^2	square kilometer
km/h	kilometers per hour
kPa	kilopascal
kt	kilotonne
kV	kilovolt
kWh	kilowatt hour
l	litre
m	meter
m^2	square meter
m^3	cubic meter
mm^3	million cubic meters
mm	millimeter
mg	milligram
mol	mole
mSv	millisievert
μg	microgram
ng	nanogram
ppm	parts per million
ppb	parts per billion
ppt	parts per trillion
s	second
SIC	Standard Industrial Classification
t	tonne
TJ	terajoule
W	watt

Appendix 6 : Geological Time

ERA	PERIOD	MILLION YEARS BEFORE PRESENT	SIGNIFICANTS EVENTS
Cenozoic			
	Quaternary	0–1.6	
	Tertiary	1.6–66.4	
Mesozoic			
	Cretaceous	66.4–144	
	Jurassic	144–208	First birds
	Triassic	208–245	First dinosaurs and mammals
Late Paleozoic			
	Permian	245–286	
	Carboniferous	286–360	First reptiles
	Devonian	360–408	First insects
Early Paleozoic			
	Silurian	408–438	First land plants
	Ordovician	438–505	First fishes
	Cambrian	505–570	First shellfish and corals
Precambrian Time			
	Proterozoic Era	544–2,500	
	Archean Era	2,500–3,800	
	Hadean Time	3,800–4,550	

The Carboniferous period is sometimes split into the
Pennsylvanian Period (286–320 million years before present) and
the Mississippian Period (320–360 million years before present).

APPENDIX 6 (cont'd). Subdivisions of the Cenozoic Era

PERIOD	EPOCH	THOUSAND YEARS BEFORE PRESENT	SIGNIFICANTS EVENTS
Quaternary			
	Holocene	0–10	
	Pleistocene	11–1,600	First modern humans
Tertiary			
	Pliocene	1,600–5,300	
	Miocene	5,300–23,700	
	Oligocene	23,700–36,600	
	Eocene	36,600–57,800	
	Paleocene	57,800–66,400	

Improved dating techniques and further study of stratigraphic sequences and fossils will require the adjustment of the periods and epochs given in Appendix 6.

For further information, consult the University of California (Berkeley) Museum of Paleontology's *Web Geological Time Machine* at http://www.ucmp.berkeley.edu/help/timeform.html. For a review of the concept of geologic time and how it is measured, see the USGS publication *Geologic Time* at http://pubs.usgs.gov/gip/geotime/.

Appendix 7 : Selected Environmental Treaties and Conventions

ANTARCTIC

Name: *Agreed Measures for the Conservation of Antarctic Fauna and Flora*
Initiated: 2 June 1964 (Brussels, Belgium)
Ratified: 1 November 1982
Description: An agreement to implement the principles and purposes of the Antarctic Treaty.
FURTHER INFORMATION: sedac.ciesin.org/pidb/
                                              www.ecolex.org

Name: *The Antarctic Treaty*
Initiated: 1 December 1959 (Washington, USA)
Ratified: 23 June 1961
Description: A treaty first signed in 1959 that encourages scientific research in the Antarctic in the absence of territorial and sovereignty claims by nations.
FURTHER INFORMATION: Vogler, J. (2000)
                                              sedac.ciesin.org/pidb/
                                              www.ecolex.org

Name: *Convention on the Conservation of Antarctic Marine Living Resources*
Initiated: 20 May 1980 (Canberra, Australia)
Ratified: 7 April 1982
Description: An agreement designed to protect the marine resources in the Antarctic Convergence and, in particular, to offer a sanctuary to whales in Antarctic waters.
FURTHER INFORMATION: Vogler, J. (2000)
                                              sedac.ciesin.org/pidb/
                                              www.ecolex.org

Name: *Convention on the Conservation of Antarctic Seals*
Initiated: 1 June 1972 (London, UK)
Ratified: 11 March 1978
Description: An agreement designed to conserve and ensure the sustainable use of seals in the Antarctic ecosystems.
FURTHER INFORMATION: Vogler, J. (2000)
                       sedac.ciesin.org/pidb/
                       www.ecolex.org

Name: *Convention on the Regulation of Antarctic Mineral Resource Activities*
Initiated: 2 June 1988 (Wellington, New Zealand)
Ratified: -----
Description: An agreement designed to regulate mining within the Antarctic.
FURTHER INFORMATION: Vogler, J. (2000)
                       sedac.ciesin.org/pidb/

Name: *Protocol to the Antarctic Treaty on Environmental Protection (PREP)*
Initiated: 4 October 1991 (Madrid, Spain)
Ratified: 14 January 1998
Description: An agreement designed to bring together the various Antarctic agreements on natural resource use and ensure comprehensive conservation of the Antarctic environment.
FURTHER INFORMATION: Vogler, J. (2000)
                       sedac.ciesin.org/pidb/
                       www.ecolex.org

ATMOSPHERE

Name: *Adjustments and Amendment to the Montreal Protocol on Substances that Deplete the Ozone Layer (London Amendment)*
Initiated: 23 June 1990 (London, UK)
Ratified: 1 January 1992

Description: A 1990 amendment to the 1987 Montreal Protocol which accelerated the phase-out reductions on the production of chlorofluorocarbons (CFCs) that destroy stratospheric ozone.
FURTHER INFORMATION: Vogler, J. (2000)
                        sedac.ciesin.org/pidb/

Name: *Amendment to the Montreal Protocol on Substances that Deplete the Ozone Layer (Copenhagen Amendment)*
Initiated: 25 November 1992 (Copenhagen, Denmark)
Ratified: 1 January 1994
Description: A 1992 amendment to the 1987 Montreal Protocol that accelerated the phase out of production of chlorofluorocarbons (CFCs) which breakdown stratospheric ozone.
FURTHER INFORMATION: Vogler, J. (2000)
                        sedac.ciesin.org/pidb/
                        www.ecolex.org

Name: *Framework Convention on Climate Change (FCCC)*
Initiated: 9 May 1992 (New York, USA)
Ratified: 21 March 1994
Description: A convention signed by over 160 countries at the 1992 Earth Summit in Rio de Janeiro, and which came into force in March 1994 following its ratification by over 50 nations. Annex I countries to the convention committed themselves to return their greenhouse gas emissions to 1990 levels by the year 2000. This commitment was modified in the 1997 Kyoto Protocol. The ultimate objective of the FCCC is to stabilize greenhouse gas concentrations at a level that would prevent dangerous anthropogenic interference with the climate system.
FURTHER INFORMATION: Vogler, J. (2000)
                        sedac.ciesin.org/pidb/
                        www.ecolex.org
                        www.unfccc.de/

Name: *The Montreal Protocol (on Substances that Deplete the Ozone Layer)*
Initiated: 16 September 1987 (Montreal, Quebec)
Ratified: 1 January 1989
Description: An agreement first signed in 1987 in Montreal, Canada, under which signatories agreed to reduce by half their production of ozone-damaging chlorofluorocarbons (CFCs) by the year 2000, and that

continued the process begun in 1985 with the Vienna Convention for the Protection of the Ozone Layer. The Protocol was subsequently strengthened in 1990, and again in 1992, with the London and Copenhagen Amendments under which all CFC production would stop by 2000.

FURTHER INFORMATION: Organisation for Economic Co-operation and Development (1999)
Vogler, J. (2000)
sedac.ciesin.org/pidb/
www.ecolex.org
www.montrealprotocol.org/
www.unep.ch/ozone/ratif.htm

Name: *Treaty Banning Nuclear Weapon Tests in the Atmosphere, in Outer Space, and Under Water*
Initiated: 5 August 1963 (Washington, USA)
Ratified: 10 October 1963
Description: An agreement to end the testing of nuclear weapons, except in underground tests, that was originally signed by the USA, the United Kingdom and the Soviet Union.
FURTHER INFORMATION: sedac.ciesin.org/pidb/
www.ecolex.org

Name: *US–Canada Air Quality Agreement (Agreement between the Government of Canada and the Government of the United States of America on Air Quality)*
Initiated: 13 March 1991 (Ottawa, Ontario)
Ratified:
Description: A 1991 agreement that committed the two countries to reductions in sulfur dioxide and nitrogen oxides emissions, and provided a forum for addressing transboundary air issues.
FURTHER INFORMATION: sedac.ciesin.org/pidb/

Name: *Vienna Convention for the Protection of the Ozone Layer*
Initiated: 22 March 1985 (Vienna, Austria)
Ratified: 22 September 1988
Description: An agreement between 20 countries in 1985 that committed them to address what, at the time, was perceived to be a

potential problem of stratospheric ozone depletion due to the emissions of chlorofluorocarbons (CFCs).

FURTHER INFORMATION: Vogler, J. (2000)
                        sedac.ciesin.org/pidb/
                        www.ecolex.org

CULTURE

Name: *Convention for the Protection of the World Cultural and Natural Heritage*
Initiated: 16 November 1972 (Paris, France)
Ratified: 17 December 1975
Description: An agreement designed to protect cultural and natural heritage sites of world significance.
FURTHER INFORMATION: sedac.ciesin.org/pidb/
                        www.ecolex.org

Name: *Declaration of the United Nations Conference on the Human Environment (UNCHE)*
Initiated: 5 June 1972 (Stockholm, Sweden)
Ratified:
Description: A declaration with 26 principles that sets an agenda for sustainable development and the environment.
FURTHER INFORMATION: Vogler, J. (2000)
                        www.biblebelievers.org.au/gc1972.htm

ENVIRONMENT and TRADE

Name: *The Marrakesh Agreement Establishing the World Trade Organization*
Initiated: April 1994
Ratified: April 1994
Description: One of the agreements reached at the conclusion of the round of talks that started in Uruguay in 1986 under the auspices of GATT. The agreement established the framework for the WTO which embodied the series of rules reached under the previous GATT agreements and also contained a preamble recognizing the need to ensure

sustainable development and environmental protection. A statement released by attendees at Marrakesh, called *The Decision on Trade and Environment*, states "There should not be, nor need be, any policy contradiction between upholding and safeguarding an open, non-discriminatory and equitable multilateral trading system on the one hand, and acting for the protection of the environment, and the promotion of sustainable development on the other."

FURTHER INFORMATION: www.wto.org/english/tratop_e/
envir_e/hist2_e.htm

Name: *The North American Agreement on Environmental Co-operation (NAAEC) (side agreement to NAFTA)*
Initiated: 13 September 1993
Ratified: 1 January 1994
Description: An agreement that creates a framework to better conserve, protect and enhance the North American environment through co-operation and effective enforcement of environmental laws.
FURTHER INFORMATION: sedac.ciesin.org/pidb/
www.ecolex.org
www.dfait-maeci.gc.ca/nafta-alena/

Name: *North American Free Trade Agreement (NAFTA)*
Initiated: 17 December 1992 (Ottawa, Ontario)
Ratified: 1 January 1994
Description: An agreement signed between the USA, Canada and Mexico in 1992 to promote free trade among the three countries. The agreement includes various environmental clauses and commits the signatories to promote sustainable development.
FURTHER INFORMATION: sedac.ciesin.org/pidb/
www.ecolex.org
www.dfait-maeci.gc.ca/nafta-alena/

Name: *Uruguay Round*
Initiated: Not applicable
Ratified: Not applicable
Description: The last of eight rounds of talks held under the auspices of GATT that focused on the removal of tariffs, the development of more liberalized trade rules, and the formation of the World Trade Organization.

FORESTRY

Name: *The Helsinki Process*
Initiated: Not applicable
Ratified: Not applicable
Description: A meeting between European countries, called the Second Ministerial Conference on the Protection of Forests in Europe, held in Helsinki in 1993, which identified the general guidelines for sustainable forestry management in Europe, and which led to the subsequent development of a series of criteria and indicators in later meetings.
FURTHER INFORMATION: www.iisd.ca/linkages/forestry/hel.html

Name: *International Tropical Timber Agreement*
Initiated: 18 November 1983 (Geneva, Switzerland)
Ratified: 1 April 1985
Description: An agreement to ensure consultation and co-operation in the production, harvesting and consumption of tropical timber and products and to ensure the sustainable use of tropical forests.
FURTHER INFORMATION: sedac.ciesin.org/pidb/
                                              www.ecolex.org

Name: *International Tropical Timber Agreement, 1994*
Initiated: 26 January 1994 (Geneva, Switzerland)
Ratified: 1 January 1997
Description: An agreement by producing and consuming nations of tropical timber to ensure that, by the year 2000, exports of tropical timber only originate from sustainably managed forests.
FURTHER INFORMATION: sedac.ciesin.org/pidb/
                                              www.ecolex.org

Name: *The Montreal Process*
Initiated: Not applicable
Ratified: Not applicable
Description: An initiative to develop a series of criteria and indicators for sustainable forest management for temperate forests outside of Europe.
FURTHER INFORMATION: www.iisd.ca/linkages/forestry/mont.html

MARINE and WATER RESOURCES

Name: *78 Agreement Between The United States and Canada on Great Lakes Water Quality (As Amended through 16 October 1983)*
Initiated: 22 November 1978 (Ottawa, Ontario)
Ratified:
Description: A 1978 accord that updated a 1972 agreement between the USA and Canada to reduce the pollution entering the Great Lakes (Superior, Michigan, Huron, Erie and Ontario) that are shared between the two countries.
FURTHER INFORMATION: sedac.ciesin.org/pidb/
www.ijc.org

Name: *Agreement Concerning Interim Arrangements Relating to Polymetallic Nodules of the Deep Sea Bed*
Initiated: 2 September 1982 (Washington, USA)
Ratified: 2 September 1982
Description: An agreement designed to avoid conflicts and overlaps of mining activities on the deep sea floor.
FURTHER INFORMATION: sedac.ciesin.org/pidb/
www.ecolex.org

Name: *Agreement for the Implementation of the Provisions of the United Nations Convention on the Law of the Sea Relating to the Conservation and Management of Straddling Fish Stocks and Highly Migratory Fish Stocks*
Initiated: 4 December 1995 (New York, USA)
Ratified: -----
Description: An agreement designed to secure enhanced management of high seas resources through provisions of monitoring under regional organizations, and the ability to adjudicate disputes under a variety of international arena.
FURTHER INFORMATION: sedac.ciesin.org/pidb/
www.ecolex.org

Name: *Agreement Relating to the Implementation of Part XI of the United Nations Convention on the Law of the Sea of 10 December 1982*
Initiated: 28 July 1994 (New York, USA)
Ratified: 16 November 1994

Description: See United Nations Convention of the Law of the Sea.
FURTHER INFORMATION: sedac.ciesin.org/pidb
                          www.ecolex.org

Name: *The Control of Eutrophication of Waters (OECD)*
Initiated: 14 November 1974
Ratified:
Description: An agreement to ensure the protection of the quality of water resources in member countries of the Organization for Economic Cooperation and Development (OECD).
FURTHER INFORMATION: sedac.ciesin.org/pidb/

Name: *Convention concerning Fishing in the Black Sea*
Initiated: 7 July 1959 (Varna, Bulgaria)
Ratified: 21 March 1960
Description: An agreement between Bulgaria, Romania and the Soviet Union to rationally utilize the fishery resources of the Black Sea.
FURTHER INFORMATION: sedac.ciesin.org/pidb/
                          www.ecolex.org

Name: *Convention for the Establishment of an Inter American Tropical Tuna Commission*
Initiated: 31 May 1949 (Washington, USA)
Ratified: 3 March 1950
Description: An agreement to establish a permanent commission to help ensure the sustainability of yellow fin and skipjack tuna. The Commission members include the USA, Canada, Costa Rica, Ecuador, France, Japan, Mexico, Nicaragua, Panama and Vanuatu.
FURTHER INFORMATION: sedac.ciesin.org/pidb/
                          www.ecolex.org

Name: *Convention for the Prohibition of Fishing with Long Driftnets in the South Pacific*
Initiated: 24 November 1989 (Wellington, New Zealand)
Ratified: 17 May 1991
Description: An agreement to prevent the use of drift nets in the South Pacific.
FURTHER INFORMATION: sedac.ciesin.org/pidb/
                          www.ecolex.org

Name: *Convention for the Protection and Development of the Marine Environment of the Wider Caribbean Region*
Initiated: 24 March 1983 (Cartagena de Indias, Colombia)
Ratified: 30 March 1986
Description: An agreement to ensure the protection and sustainable management of the marine environment in the Caribbean region.
FURTHER INFORMATION: sedac.ciesin.org/pidb/
www.ecolex.org
www.cep.unep.org/

Name: *Convention on Fishing and Conservation of the Living Resources of the High Seas*
Initiated: 29 April 1958 (Geneva, Switzerland)
Ratified: 20 March 1966
Description: An agreement to help ensure the sustainable use of marine resources.
FURTHER INFORMATION: sedac.ciesin.org/pidb/
www.ecolex.org

Name: *Convention on the High Seas*
Initiated: 29 April 1958 (Geneva, Switzerland)
Ratified: 30 September 1962
Description: An agreement to codify rules on international law on the high seas.
FURTHER INFORMATION: sedac.ciesin.org/pidb/
www.ecolex.org

Name: *Convention on Wetlands of International Importance especially as Waterfowl Habitat (The Ramsar Convention on Wetlands)*
Initiated: 2 February 1971 (Ramsar, Iran)
Ratified: 21 December 1975
Description: An agreement by 122 countries to facilitate the conservation and wise use of wetlands.
FURTHER INFORMATION: sedac.ciesin.org/pidb/
www.ecolex.org
www.ramsar.org

Name: *International Convention for the Regulation of Whaling (International Whaling Convention)*
Initiated: 2 December 1946 (Washington, USA)
Ratified: 10 November 1948
Description: A convention signed in 1946 that established the International Whaling Commission, mandated to preserve whale stocks. The Commission has been highly controversial and failed to prevent the near extinction of several whale species. More recently, a moratorium on commercial whaling has helped some species to increase in number but controversy still remains, as some members of the commission would like to ban all commercial whaling in perpetuity while other countries would like to resume commercial whaling.
FURTHER INFORMATION: Vogler, J. (2000)
                        sedac.ciesin.org/pidb/
                        www.ecolex.org

Name: *United Nations Convention on the Law of the Sea*
Initiated: 10 December 1982 (Montego Bay, Jamaica)
Ratified: 16 November 1994
Description: A convention developed by a series of international meetings in the 1970s and 1980s to govern the exploitation of marine resources. The Convention established, among other things, 200 nautical mile exclusive economic zones (EEZs) from the coastline of maritime states.
FURTHER INFORMATION: Vogler, J. (2000)
                        sedac.ciesin.org/pidb/
                        www.ecolex.org

POLLUTION

Name: *Agreement Between the Government of the United States of America and the Government of Canada Concerning the Transboundary Movement of Hazardous Wastes*
Initiated: 28 October 1986 (Ottawa, Ontario)
Ratified: 8 November 1986
Description: An agreement between the USA and Canada on the appropriate treatment and transport of hazardous wastes.
FURTHER INFORMATION: sedac.ciesin.org/pidb/

Name: *Agreement Concerning the Protection of the Waters of the Mediterranean Shores*
Initiated: 10 May 1976 (Monaco)
Ratified: 1 January 1981
Description: An agreement between France, Monaco and Italy to ensure co-operation and co-ordination of policies to improve the coastal waters of the three nations.
FURTHER INFORMATION: sedac.ciesin.org/pidb/
                         www.ecolex.org

Name: *Bamako Convention on the Ban of the Import into Africa and the Control of Transboundary Movement and Management of Hazardous Wastes Within Africa*
Initiated: 30 January 1991 (Bamako, Mali)
Ratified: 22 April 1998
Description: An agreement to prevent the dumping or improper disposal of hazardous wastes in Africa.
FURTHER INFORMATION: sedac.ciesin.org/pidb/
                         www.ecolex.org

Name: *Basel Convention on the Control of Transboundary Movements of Hazardous Wastes and their Disposal*
Initiated: 22 March 1989 (Basel, Switzerland)
Ratified: 05 May 1992
Description: A 1992 convention, commonly referred to as the Basel Convention, which ultimately led to an agreement to phase out the dumping and inappropriate disposal of hazardous wastes in developing countries.
FURTHER INFORMATION: Organization for Economic Co-operation
                         and Development (1999)
                         sedac.ciesin.org/pidb/
                         www.ecolex.org
                         www.unep.ch/basel/index.html

Name: *Convention Concerning Safety in the Use of Asbestos*
Initiated: 24 June 1986 (Geneva, Switzerland)
Ratified: 16 June 1986
Description: An agreement to prevent the potential health hazards associated with the use of asbestos.

FURTHER INFORMATION: sedac.ciesin.org/pidb/
www.ecolex.org

Name: *Convention for the Protection of the Mediterranean Sea against Pollution*
Initiated: 16 February 1976 (Barcelona, Spain)
Ratified: 12 February 1978
Description: An agreement designed to help protect the Mediterranean marine environment.
FURTHER INFORMATION: sedac.ciesin.org/pidb/
www.ecolex.org
www.unepmap.gr/

Name: *Convention for the Protection of the Rhine against Chemical Pollution*
Initiated: 3 December 1976 (Bonn, Germany)
Ratified: 1 March 1983
Description: An agreement between countries bordering on the Rhine River to help prevent chemical spills and pollution.
FURTHER INFORMATION: sedac.ciesin.org/pidb/
www.ecolex.org

Name: *Convention on Environmental Impact Assessment in a Transboundary Context*
Initiated: 25 February 1991 (Espoo, Finland)
Ratified: 10 September 1997
Description: An agreement to use Environmental Impact Assessments so as to prevent transboundary deterioration of the environment.
FURTHER INFORMATION: sedac.ciesin.org/pidb/
www.ecolex.org

Name: *Convention on Long-Range Transboundary Air Pollution*
Initiated: 13 November 1979 (Geneva, Switzerland)
Ratified: 16 March 1983
Description: An agreement to reduce long-range transboundary air pollution.
FURTHER INFORMATION: sedac.ciesin.org/pidb/
www.ecolex.org

Name: *Convention on the Prevention of Marine Pollution by Dumping Wastes and Other Matter (London Dumping Convention)*
Initiated: 29 December 1972 (London, UK)
Ratified: 30 August 1975
Description: First signed in 1972, and ratified by 75 countries, the Convention prohibits the dumping of specified wastes at sea, including radioactive wastes.
FURTHER INFORMATION: Vogler, J. (2000)
                                              sedac.ciesin.org/pidb/
                                              www.ecolex.org

Name: *Convention on the Prevention of Marine Pollution from Land-Based Sources*
Initiated: 4 June 1974 (Paris, France)
Ratified: 6 May 1978
Description: An agreement designed to prevent land-based pollution of the marine environment.
FURTHER INFORMATION: sedac.ciesin.org/pidb/

Name: *Convention on the Prohibition of the Development, Production and Stockpiling of Bacteriological (Biological) and Toxic Weapons and on their Destruction*
Initiated: 10 April 1972
Ratified: 26 March 1975
Description: An agreement designed to prevent the production and storage of chemical weapons.
FURTHER INFORMATION: sedac.ciesin.org/pidb/
                                              www.ecolex.org

Name: *European Agreement on the Restriction of the Use of Certain Detergents in Washing and Cleaning Products*
Initiated: 16 September 1968 (Strasbourg, France)
Ratified: 16 February 1971
Description: An agreement designed to protect freshwater environments, particularly for recreational and wildlife purposes, by controlling the use of detergents.
FURTHER INFORMATION: sedac.ciesin.org/pidb/
                                              www.ecolex.org

Name: *International Convention on Civil Liability for Oil Pollution Damage*
Initiated: 29 November 1969 (Brussels, Belgium)
Ratified: 19 June 1975
Description: An agreement designed to standardize the rules of compensation for oil spills.
FURTHER INFORMATION: sedac.ciesin.org/pidb/
www.ecolex.org

Name: *International Convention on the Establishment of an International Fund for Compensation for Oil Pollution Damage*
Initiated: 18 December 1971 (Brussels, Belgium)
Ratified: 16 October 1978
Description: An amendment to a previous agreement that helps ensure that the owners of oil cargoes bear a responsibility for paying compensation for oil spills.
FURTHER INFORMATION: sedac.ciesin.org/pidb/

Name: *International Convention for the Prevention of Pollution from Ships (MARPOL)*
Initiated: 2 November 1973 (London, UK)
Ratified: ---
Description: An agreement that regulates the discharge of petroleum products and other pollutants at sea.
FURTHER INFORMATION: Vogler, J. (2000)
sedac.ciesin.org/pidb/
www.ecolex.org

Name: *Kyoto Protocol*
Initiated: 11 December 1997 (Kyoto, Japan)
Ratified: ---
Description: The 1997 agreement by parties to the framework convention on climate change which sets binding emission constraints on Annex I (listed as Annex B countries in the Protocol) for the period 2008–12. Collectively, Annex B countries are committed to a 5 percent reduction in greenhouse gas emissions by 2008–12 relative to base levels in 1990. Some Annex B countries (such as Australia) secured increases in emissions over 1990 levels, some agreed to meet 1990 levels (such as New Zealand) and some agreed to exceed the 5 percent reduction (such as Germany). The Protocol allows for the possibility of

carbon sinks and emissions trading although the precise rules are to be determined upon at a later date.

FURTHER INFORMATION: Vogler, J. (2000)

www.ecolex.org

www.cop4.org/kp/kp.html

Name: *Protocol Concerning Marine Pollution Resulting from Exploration and Exploitation of the Continental Shelf*

Initiated: 29 March 1989 (Kuwait)

Ratified: 17 February 1990

Description: An agreement to help prevent marine pollution from exploration and exploitation of the sea-bed.

FURTHER INFORMATION: sedac.ciesin.org/pidb/

www.ecolex.org

Name: *Protocol for the Prevention of Pollution of the Mediterranean Sea by Dumping from Ships and Aircraft*

Initiated: 16 February 1976 (Barcelona, Spain)

Ratified: 12 February 1978

Description: An agreement to help prevent the disposal of wastes in the Mediterranean sea.

FURTHER INFORMATION: sedac.ciesin.org/pidb/

www.ecolex.org

Name: *Protocol for the Protection of the Mediterranean Sea against Pollution from Land-Based Sources*

Initiated: 17 May 1980 (Athens, Greece)

Ratified: 17 June 1983

Description: An agreement to prevent pollution in the Mediterranean from various land-based sources.

FURTHER INFORMATION: sedac.ciesin.org/pidb/

www.ecolex.org

Name: *Protocol Relating to Intervention on the High Seas in Cases of Pollution by Substances other than Oil*

Initiated: 2 November 1973 (London, UK)

Ratified: 30 March 1983

Description: An agreement that allows coastal states to act on the high seas to prevent pollution, other than oil.

FURTHER INFORMATION: sedac.ciesin.org/pidb/
www.ecolex.org

Name: *Protocol to the 1979 Convention on Long-Range Transboundary Air Pollution Concerning the Control of Emissions of Volatile Organic Compounds or Their Transboundary Fluxes*
Initiated: 18 November 1991 (Geneva, Switzerland)
Ratified: 29 September 1997
Description: An agreement to reduce the emissions of volatile organic compounds.
FURTHER INFORMATION: sedac.ciesin.org/pidb/
www.ecolex.org
www.unece.org/env/lrtap/

Name: *Protocol to the 1979 Convention on Long-Range Transboundary Air Pollution on Further Reduction of Sulphur Emissions*
Initiated: 14 June 1994 (Oslo, Norway)
Ratified: 5 August 1998
Description: An agreement to reduce the emissions of sulfur dioxide and their long-range transportation.
FURTHER INFORMATION: www.ecolex.org
www.unece.org/env/lrtap/

Name: *Protocol to the 1979 Convention on Long-Range Transboundary Air Pollution on Long-Term Financing of Co-operative Program for Monitoring and Evaluation of the Long-Term Transmission of Air Pollutants in Europe*
Initiated: 28 February 1984 (Geneva, Switzerland)
Ratified: 28 January 1988
Description: An agreement on the monitoring of long-range pollutants in
Europe.
FURTHER INFORMATION: sedac.ciesin.org/pidb/
www.ecolex.org
www.unece.org/env/lrtap/

Name: *Protocol to the 1979 Convention on Long-Range Transboundary Air Pollution on Persistent Organic Pollutants*
Initiated: 24 June 1998 (Aarhus, Denmark)
Ratified: -----

Description: An agreement to reduce the emissions of persistent organic pollutants and their long-range transportation.
FURTHER INFORMATION: www.ecolex.org
www.unece.org/env/lrtap/

Name: *Protocol to the 1979 Convention on Long-Range Transboundary Air Pollution on the Reduction of Sulphur Emissions or their Transboundary Fluxes by at Least 30 Percent*
Initiated: 8 July 1985 (Helsinki, Finland)
Ratified: 2 September 1987
Description: An agreement by some European countries to reduce sulfur dioxide emissions by 30 percent by 1993.
FURTHER INFORMATION: sedac.ciesin.org/pidb/
www.ecolex.org
www.unece.org/env/lrtap/

Name: *Protocol to the International Convention on the Establishment of an International Fund of Compensation for Oil Pollution Damage*
Initiated: 19 November 1976 (London, UK)
Ratified: 22 November 1994
Description: An agreement to establish international compensation for the effects of oil pollution damages.
FURTHER INFORMATION: sedac.ciesin.org/pidb/
www.ecolex.org

Name: *Rotterdam Convention*
Initiated: 10 September 1998 (Rotterdam, the Netherlands)
Ratified: -----
Description: The Rotterdam Convention lists a set of procedures for providing adequate information to importing countries about potentially hazardous internationally traded materials.
FURTHER INFORMATION: www.ecolex.org
www.fao.org/waicent/FAOINFO/AGRI
CULT/AGP/AGPP/Pesticid/PIC/dipcon.htm

SUSTAINABLE DEVELOPMENT

Name: *Convention on Biological Diversity*
Initiated: 22 May 1992 (Rio de Janeiro, Brazil)
Ratified: 29 December 1993
Description: The 1992 convention on biological diversity which allows for the free trade of genetic resources while providing ways for rich countries to finance biodiversity conservation.
FURTHER INFORMATION: sedac.ciesin.org/pidb/
                                 www.ecolex.org
                                 www.biodiv.org

Name: *International Convention to Combat Desertification in those Countries Experiencing Serious Drought and/or Desertification, Particularly in Africa*
Initiated: 14 October 1994 (Paris, France)
Ratified: 26 December 1996
Description: An agreement to help countries facilitate plans to address the problems of desertification within their countries, as well as to encourage technology transfer to help countries overcome the deleterious effects of desertification.
FURTHER INFORMATION: sedac.ciesin.org/pidb/
                                 www.ecolex.org
                                 www.iisd.ca/linkages

Name: *The Rio Declaration on Environment and Development*
Initiated: 13 June 1992 (Rio de Janeiro, Brazil)
Ratified: 13 June 1992
Description: A declaration of 27 principles issued at the 1992 United Nations Conference on Environment and Development, including the principle of sustainable development.
FURTHER INFORMATION: Vogler, J. (2000)
                                 sedac.ciesin.org/pidb/

Name: *Treaty for Amazonian Cooperation*
Initiated: 3 July 1978 (Brasilia, Brazil)
Ratified: 2 August 1980
Description: An agreement by Bolivia, Brazil, Colombia, Ecuador, Guyana, Peru, Suriname and Venezuela to ensure the sustainable

development of the Amazon region and an equitable division of its benefits.

FURTHER INFORMATION: sedac.ciesin.org/pidb/
                                             www.ecolex.org

WILDLIFE

Name: *African Convention on the Conservation of Nature and Natural Resources*
Initiated: 15 September 1968 (Algiers, Algeria)
Ratified: 16 June 1969
Description: An agreement to promote the conservation, utilization and development of the soil, water, plant and wildlife resources of the African continent.
FURTHER INFORMATION: sedac.ciesin.org/pidb/
                                             www.ecolex.org

Name: *Agreement on Conservation of Polar Bears*
Initiated: 15 November 1973 (Oslo, Norway)
Ratified: 26 May 1976
Description: An agreement to protect the polar bear through appropriate management measures.
FURTHER INFORMATION: sedac.ciesin.org/pidb/
                                             www.ecolex.org

Name: *Agreement for the Establishment of a General Fisheries Council for the Mediterranean*
Initiated: 24 September 1949 (Rome, Italy)
Ratified: 3 December 1963
Description: An agreement to help achieve international co-operation in the use of the fisheries in the Mediterranean.
FURTHER INFORMATION: sedac.ciesin.org/pidb/
                                             www.ecolex.org

Name: *ASEAN Agreement on the Conservation of Nature and Natural Resources*
Initiated: 9 July 1985 (Kuala Lumpur, Malaysia)
Ratified: -----

Description: An agreement among members of the Association of Southeast Asian nations to jointly and separately help conserve their natural capital.
FURTHER INFORMATION: sedac.ciesin.org/pidb/
www.ecolex.org

Name: *Benelux Convention on Nature Conservation and Landscape Protection*
Initiated: 8 June 1982 (Brussels, Belgium)
Ratified: 1 October 1983
Description: An agreement between the Netherlands, Luxembourg and Belgium to co-operate in the management of their natural heritage.
FURTHER INFORMATION: sedac.ciesin.org/pidb/
www.ecolex.org

Name: *Biosafety Protocol*
Initiated: 29 January 2000 (Cartagena, Spain)
Ratified: (will remain open for signature from 5 June 2000 to 4 June 2001)
Description: Formally known as the Cartagena Protocol on Biosafety, that was signed by the Conference of the Parties to the Convention on Biological Diversity in May 2000. The Protocol seeks to regulate biodiversity via the regulation of the international transport of living genetically modified organisms, and also to assist in the protection of traditional knowledge.
FURTHER INFORMATION:
www.biodiv.org/biosafe/Protocol/Index.html

Name: *Convention on Nature Protection and Wildlife Preservation in the Western Hemisphere*
Initiated: 12 October 1940 (Washington, USA)
Ratified: 1 May 1942
Description: A Pan-American agreement to help conserve the flora and fauna in the western hemisphere.
FURTHER INFORMATION: sedac.ciesin.org/pidb/
www.ecolex.org

Name: *Convention on the African Migratory Locust*
Initiated: 25 May 1962 (Kano, Nigeria)

Ratified: 25 July 1968
Description: An extension to a 1952 agreement signed by 22 African countries to include the control of all other migratory species locusts in Africa.
FURTHER INFORMATION: sedac.ciesin.org/pidb/
www.ecolex.org

Name: *Convention on the Conservation of European Wildlife and Natural Habitats*
Initiated: 19 September 1979 (Berne, Switzerland)
Ratified: 1 June 1982
Description: An agreement by 21 European countries and Burkino Faso and Senegal to co-operate to conserve European wildlife.
FURTHER INFORMATION: sedac.ciesin.org/pidb/
www.ecolex.org

Name: *Convention on the Conservation of Migratory Species of Wild Animals, as amended 1985*
Initiated: 23 June 1979 (Bonn, Germany)
Ratified: 1 November 1983
Description: An agreement by 39 countries to protect migratory wildlife species.
FURTHER INFORMATION: sedac.ciesin.org/pidb/
www.ecolex.org

Name: *Convention on the International Trade in Endangered Species of Wild Flora and Fauna (CITES)*
Initiated: 3 March 1973 (Washington, USA)
Ratified: 1 July 1975
Description: A convention initiated in 1973 that seeks to protect endangered plant and animal species from extinction. CITES has been signed by over 100 countries and has been an important mechanism in preventing (or at least restricting) the trade in animal parts, such as ivory.
FURTHER INFORMATION: Organization for Economic Co-operation and Development (1999)
sedac.ciesin.org/pidb/
www.ecolex.org
www.cites.org/

Name: *Convention Relative to the Preservation of Fauna and Flora in their Natural State*
Initiated: 8 November 1933 (London, UK)
Ratified: 14 January 1936
Description: An agreement by 11 countries and dominions to protect flora and fauna, particularly in Africa, through the control of the trade of trophies, restrictions of certain types of hunting and protection of listed species.
FURTHER INFORMATION: sedac.ciesin.org/pidb/
                        www.ecolex.org

Name: *European Convention for the Protection of Animals During International Transport*
Initiated: 13 December 1968 (Paris, France)
Ratified: 20 February 1971
Description: An agreement by 22 European countries to prevent the suffering of animals during transportation across their borders.
FURTHER INFORMATION: sedac.ciesin.org/pidb/
                        www.ecolex.org

Name: *International Convention for the Protection of Birds*
Initiated: 18 October 1950 (Paris, France)
Ratified: 17 January 1963
Description: An agreement to protect birds, especially migratory birds and those species in danger of extinction. The convention includes articles about when birds should not be hunted.
FURTHER INFORMATION: sedac.ciesin.org/pidb/
                        www.ecolex.org

Name: *International Plant Protection Convention*
Initiated: 6 December 1951 (Rome, Italy)
Ratified: 3 April 1952
Description: An agreement to prevent the spread of plant diseases and to co-ordinate measures for preventing pests moving between the signatory countries.
FURTHER INFORMATION: sedac.ciesin.org/pidb/
                        www.ecolex.org

Name: *Plant Protection Agreement for the Asia and Pacific Region*
Initiated: 27 February 1956 (Rome, Italy)
Ratified: 2 July 1956
Description: An agreement between 26 countries to prevent the spread of plant diseases in the Asian and Pacific regions by certification of the trade of plants, and special measures to prevent the introduction of South American leaf blight to rubber trees.
FURTHER INFORMATION: sedac.ciesin.org/pidb/
www.ecolex.org

Name: *Statutes of the International Union for Conservation of Nature and Natural Resources (as amended)*
Initiated: 5 October 1948 (Fontainebleau, France)
Ratified: 5 October 1948
Description: An agreement that formed the International Union for the Conservation of Nature (IUCN) and that now includes 139 countries represented by both governments and non-governmental organizations. The mandate of the IUCN is to conserve the integrity and diversity of nature.
FURTHER INFORMATION: sedac.ciesin.org/pidb/
www.ecolex.org
www.iucn.org

Name: *World Charter for Nature (1982)*
Initiated:
Ratified:
Description: A declaration to ensure nations follow principles of conservation and ensure the functioning of natural systems.
FURTHER INFORMATION: sedac.ciesin.org/pidb/